Tensor Voting
A Perceptual Organization Approach
to Computer Vision and Machine Learning

Tensor Voting: A Perceptual Organization Approach to Computer Vision and Machine Learning
Philippos Mordohai and Gérard Medioni

ISBN: 978-3-031-01114-6   paperback
ISBN: 978-3-031-02242-5   ebook

DOI: 10.1007/978-3-031-02242-5

A Publication in the Springer Series:
SYNTHESIS LECTURES ON IMAGE, VIDEO, AND MULTIMEDIA PROCESSING
Lecture #8

Series Editor: Alan C. Bovik, University of Texas, Austin
ISSN   Print 1559-8136   Electronic 1559-8144

First Edition
10 9 8 7 6 5 4 3 2 1

Tensor Voting
A Perceptual Organization Approach
to Computer Vision and Machine Learning

Philippos Mordohai
University of North Carolina

Gérard Medioni
University of Southern California

SYNTHESIS LECTURES ON IMAGE, VIDEO, AND MULTIMEDIA PROCESSING #8

ABSTRACT

This lecture presents research on a general framework for perceptual organization that was conducted mainly at the Institute for Robotics and Intelligent Systems of the University of Southern California. It is not written as a historical recount of the work, since the sequence of the presentation is not in chronological order. It aims at presenting an approach to a wide range of problems in computer vision and machine learning that is data-driven, local and requires a minimal number of assumptions. The tensor voting framework combines these properties and provides a unified perceptual organization methodology applicable in situations that may seem heterogeneous initially. We show how several problems can be posed as the organization of the inputs into salient perceptual structures, which are inferred via tensor voting. The work presented here extends the original tensor voting framework with the addition of boundary inference capabilities, a novel re-formulation of the framework applicable to high-dimensional spaces and the development of algorithms for computer vision and machine learning problems. We show complete analysis for some problems, while we briefly outline our approach for other applications and provide pointers to relevant sources.

KEYWORDS

Perceptual organization, computer vision, machine learning, tensor voting, stereo vision, dimensionality estimation, manifold learning, function approximation, figure completion

Contents

Acknowledgements

The authors are grateful to Adit Sahasrabudhe and Matheen Siddiqui for assisting with some of the new experiments presented here and to Lily Cheng for her feedback on the manuscript. We would also like to thank Gideon Guy, Mi-Suen Lee, Chi-Keung Tang, Mircea Nicolescu, Jinman Kang, Wai-Shun Tong, and Jiaya Jia for allowing us to present some results of their research in this book.

CHAPTER 1

Introduction

The research presented here attempts to develop a general, unsupervised, data-driven methodology to address problems in computer vision and machine learning from a perceptual organization perspective. It is founded on the tensor voting framework, which in its preliminary form was proposed by Guy in [24] and was further developed by Lee [49] and Tang [103]. We include more recent developments, which can also be found in [64]. Tensor voting is a computational framework for perceptual organization based on the Gestalt principles. It has mainly been applied for organizing generic points (tokens) into coherent groups and for computer vision problems that are formulated as perceptual organization of simple tokens.

The work presented here extends the description of the book by Medioni, Tang, and Lee [60] in many ways. First, by applying tensor voting directly to images for core computer vision problems, taking into account the inherent difficulties associated with them. Second, by proposing a new N D implementation that opens the door for many applications in instance-based learning. Finally, by augmenting data representation and voting with first-order properties that allow the inference of boundaries and terminations.

1.1   MOTIVATION

The tensor voting framework attempts to implement the often conflicting Gestalt principles for perceptual organization. These principles were proposed in the first half of the twentieth century by psychologists in Central Europe. Some of the most representative research can be found in the texts of Köhler [43], Wertheimer [118], and Koffka [42]. At the core of Gestalt psychology is the axiom, "the whole is greater than the sum of the parts" [118]. In other words, configurations of simple elements give rise to the perception of more complex structures. Fig. 1.1 shows a few of the numerous factors discussed in [118].

Even though Gestalt psychologists mainly addressed grouping problems in 2D, the generalization to 3D is straightforward, since salient groupings in 3D can be detected by the human visual system based on the same principles. This is the basis of our approach to stereo vision, where the main premise is that correct pixel correspondences reconstructed in 3D form *salient* surfaces, while wrong correspondences are not well aligned and do not form any coherent structures. The term *saliency* is used in our work to indicate the quality of features to be important,

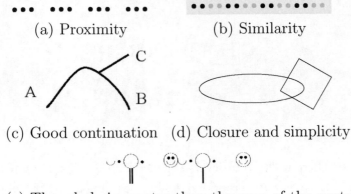

(e) The whole is greater than the sum of the parts

FIGURE 1.1: Some examples of the Gestalt principles. In (a) the dots are grouped in four groups according to proximity. In (b) the darker dots are grouped in pairs, as are the lighter ones. In (c) the most likely grouping is A to B, and not A to C, due to the smooth continuation of curve tangent from A to B. In (d), the factors of closure and simplicity generate the perception of an ellipse and a diamond. Finally, (e) illustrates that the whole is greater than the sum of the parts.

stand out conspicuously, be prominent and attract our attention. Our definition of saliency is that of Shashua and Ullman's [99] *structural saliency*, which is a product of proximity and good continuation. It is different from that of Itti and Baldi [33], where saliency is defined as the property to attract attention due to reasons that include novelty and disagreement with surrounding elements. The term *alignment* is used here to indicate good continuation and not configuration.

In our research, we are interested in inferring salient groups that adhere to the "matter is cohesive" principle of Marr [58]. For instance, given an image, taken from the Berkeley Segmentation Dataset (http://www.cs.berkeley.edu/projects/vision/grouping/) that contains texture, one can perform high-pass filtering and keep high responses of the filter as edges (Fig. 1.2). On the other hand, a human observer selects the most salient edges due to their good alignment that forms either familiar or coherent shapes, as in Fig. 1.2(c). One cannot ignore the effect of familiarity in the ability of people to detect meaningful groupings, but a machine-based perceptual organization system should be able to improve upon the performance of the high-pass filter and move toward being more similar to human performance. Significant edges are not only characterized by high responses of the filter, but, more importantly, the elementary edgels that form them are aligned with other edgels to form typically smooth, closed contours that encompass regions that are consistent in color or texture. Edgels that are not well aligned, as those in the interior of unstructured texture, are usually less important for the understanding of the scene.

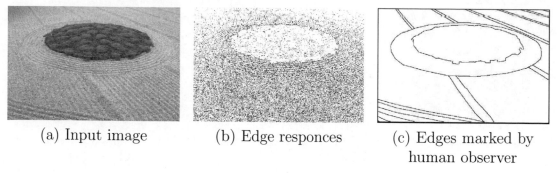

(a) Input image   (b) Edge responces   (c) Edges marked by
                                                     human observer

FIGURE 1.2: An image with texture, outputs of a simple edge detector and human-marked edges.

We are also interested in the simultaneous inference of all types of structures that may exist in a scene, as well as their boundaries and intersections. Given two images of a table from different viewpoints such as that in Fig. 1.3(a), we would like to be able to group pixel correspondences and infer the surfaces. We would also like to infer the intersections of these surfaces, which are the edges of the table. Furthermore, the intersections of these intersections are the corners of the table, which, despite their infinitesimal size, carry significant information about the configuration of the objects in the scene. The inference of integrated descriptions is a major advantage of tensor voting over competing approaches. The fact that different feature types are not independent of each other, but rather certain types occur at special configurations of other types, is an important consideration in this work.

The notion of structural saliency extends to spaces of higher dimensionality, even though it is hard to be confirmed by the human visual system. Samples in high-dimensional spaces that are produced by a consistent system, or are somehow meaningfully related, form smooth structures in the same way as point samples from a smooth surface measured with a range finder

(a) Input image   (b) Surface inter sections and corners

FIGURE 1.3: An image (potentially from a stereo pair), surface intersections, and corners.

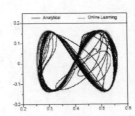
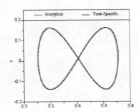

(a) Humanoid robot   (b) Time course of learning   (c) Trajectory after learning and analytical solution

FIGURE 1.4: Training a humanoid robot to draw figure "8" using an unsupervised machine learning approach [114].

provide an explicit representation of the surface. The detection of important structures and the extraction of information about the underlying process from them is the object of instance-based learning. An example from the field of kinematics taken from [114] can be seen in Fig. 1.4 where a humanoid robot tries to learn how to draw figure "8". Each observation in this example consists of 30 positions, velocities and accelerations of the joints of the arm of the robot. Therefore, each state can be represented as a point in a 90D space. Even though the analytical computation of the appropriate commands to perform the task is possible, a machine learning approach based on the premise that points on the trajectory form a low-dimensional manifold in the high-dimensional space proves to be very effective, as seen in Fig. 1.4(c) where the two solutions almost coincide. A significant contribution of [64] is a new efficient implementation of the tensor voting framework that is applicable for salient structure inference in very high-dimensional spaces and can be a powerful tool for instance-based learning.

1.2 APPROACH

Heeding the principles discussed in the previous section, we aim at the development of an approach that is both effective at each problem and also general and flexible. Tensor voting serves as the core of all the algorithms we developed, since it meets the requirements we consider necessary. It is

- local,
- data driven,
- unsupervised,
- able to process large amounts of data,
- able to represent all structure types and their interactions simultaneously,

- robust to noise,
- amenable to a least-commitment strategy, postponing the need for hard decisions.

The strength of tensor voting is due to two factors: data representation with second-order, symmetric, nonnegative definite tensors and first-order, polarity vectors; and local information propagation in the form of tensor and vector votes. The core of the representation is the second-order tensor, which encodes a saliency value for each possible type of structure along with its normal and tangent orientations. The eigenvalues and eigenvectors of the tensor, which can be conveniently expressed in matrix form, provide all the information. For instance, the eigenvalues of a 3D second-order tensor encode the saliency of a token as a surface inlier, a curve inlier or surface intersection, or as a curve intersection or volume inlier. The eigenvectors, on the other hand, correspond to the normal and tangent orientations, depending on the structure type. If the token belongs to a curve, the eigenvectors that correspond to the two largest eigenvalues are normal to the curve, while the eigenvector that corresponds to the minimum eigenvalue is tangent to it. Perceptual organization occurs by combining the information contained in the arrangement of these tokens by tensor voting. During the voting process, each token communicates its preferences for structure type and orientation to its neighbors in the form of votes, which are also tensors that are cast from token to token. Each vote has the orientation the receiver would have if the voter and receiver were part of the same smooth perceptual structure.

The major difference between tensor voting and other methodologies is the absence of an explicit objective function. In tensor voting, the solution emerges from the data and, is not enforced upon them. If the tokens align to form a curve, then the accumulation of votes will produce high curve saliency and an estimate for the tangent at each point. On the other hand, if one poses the problem as the inference of the most salient surface from the data under an optimization approach, then a surface that optimizes the selected criteria will be produced, even if it is not the most salient structure in the dataset. A simple illustration of the effectiveness of local, data-driven methods for perceptual organization can be seen in Fig. 1.5, where we are presented with unoriented point inputs and asked to infer the most likely structure. A global method such as principal component analysis (PCA) [37] can be misled by the fact that the points span a 2D subspace and fit the best surface. Tensor voting, on the other hand, examines the data locally and is able to detect that the structure is intrinsically 1D. The output is a curve which is consistent with human perception. Other hypotheses, such as whether the inputs form a surface, do not need to be formed or examined due to the absence of any prior models besides smoothness. The correct hypothesis emerges from the data. Furthermore, tensor voting allows the interaction of different types of structures, such as the intersection of a surface and a curve. To our knowledge, this is not possible with any other method.

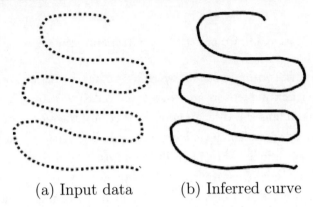

(a) Input data   (b) Inferred curve

FIGURE 1.5: Tensor voting is able to infer the correct intrinsic dimensionality of the data, which is 1D, despite the fact that it appears as 2D if observed globally. The correct perceptual structure, a curve, is inferred without having to examine potential surface hypotheses.

Additional advantages brought about by the representation and voting schemes are noise robustness and the ability to employ a least-commitment approach. As shown in numerous experiments and publications [25, 60, 104], tensor voting is robust to very large percentages of outliers in the data. This is due to the fact that random outliers cast inconsistent votes, which do not affect the solution significantly. This does not hold when there is a systematic source of errors, as is the case in many computer vision problems. Examples of such problems are shown in the appropriate chapters. Regarding the avoidance of premature decisions, the capability of the second-order tensor to encompass saliencies for all structure types allows us not having to decide whether a token is an inlier or an outlier before all the necessary information has been accumulated.

Finally, we believe that a model-free, data-driven design is more appropriate for a general approach since it is easier to generalize to new domains and more flexible in terms of the types of solutions it can infer. The main assumption in the algorithms described here is smoothness, which is a very weak and general model. Moreover, the absence of global computations increases the amount of data that can be processed since computational and storage complexity scale linearly with the number of tokens, if their density does not increase.

1.3 OUTLINE
The book is organized in three parts:

- The original tensor voting framework in 2D and 3D and its application to computer vision problems.
- Tensor voting in high dimensions with applications to machine learning problems.
- First-order voting for boundary inference and its application to figure completion.

We begin by illustrating the tensor voting framework in 2D and 3D. We show how tokens, which are simple primitives that potentially belong to perceptual structures, can be represented by second-order, symmetric, nonnegative definite tensors. Then, we present the mechanism for information propagation we have designed. The results of this propagation are structural saliency values for each potential structure type along with the preferred normal and tangent orientations at each token. Experimental results demonstrate the effectiveness of our approach, as well as its robustness to noise.

A large part of our research efforts is devoted to the development of a stereo reconstruction approach, which is presented in Chapter 3. Stereovision can be cast as a perceptual organization problem under the premise that solutions must comprise coherent structures. These structures become salient due to the alignment of potential pixel correspondences reconstructed in a 3D space. Tensor voting is performed to infer the correct matches that are generated by the true scene surfaces as inliers of smooth perceptual structures. The retained matches are grouped into smooth surfaces and inconsistent matches are rejected. Disparity hypotheses for pixels that remain unmatched are generated based on the color information of nearby surfaces and validated by ensuring the good continuation of the surfaces via tensor voting. Thus, information is propagated from more to less reliable pixels considering both geometric and color information.

A recent, major enhancement of the framework is an efficient N D implementation, which is described in Chapter 4. We present a new implementation of tensor voting that significantly reduces computational time and storage requirements, especially in high-dimensional spaces, and thus can be applied to machine learning problems, as well as a variety of new domains. This work is based on the observation that the Gestalt principles still apply in spaces of higher dimensionality. The computational and storage requirements of the original implementation prevented its wide application to problems in high dimensions. This is no longer the case with the new implementation which opens up an array of potential applications mostly in the field of instance-based learning.

Chapter 5 presents our approach to machine learning problems. We address unsupervised manifold learning from observations in high-dimensional spaces using the new efficient implementation of tensor voting. We are able to estimate local dimensionality and structure, measure geodesic distances, and perform nonlinear interpolation. We first show that we can obtain reliable dimensionality estimates at each point. Then, we present a quantitative evaluation of our results in the estimation of a local manifold structure using synthetic datasets with known ground truth. We also present results on datasets with varying dimensionality and intersections under severe noise corruption, which would have been impossible to process with current state-of-the-art methods. We also address function approximation from samples, which is an important problem with many applications in machine learning. We propose a noniterative, local, nonparametric approach that can successfully approximate nonlinear functions in high-dimensional spaces in the presence of noise. We present quantitative results on data with varying density, outliers, and perturbation, as well as real data.

In the third part, we describe the recent addition of first-order representation and voting that complement the strictly second-order previous formulation of [24, 49, 60, 103]. The augmented framework presented in Chapter 7, makes the inference of the terminations of perceptual structures possible. Polarity vectors are now associated with each token and encode the support the token receives for being a termination of a perceptual structure. The new representation exploits the essential property of boundaries to have all their neighbors, at least locally, on the same side of a half-space. The work presented in this chapter can serve as the foundation for more complex perceptual organization problems.

One such problem is addressed in Chapter 7, where we attempt to explain certain phenomena associated with figure completion within the tensor voting framework. Endpoints and junctions play a critical role in contour completion by the human visual system, and should be an integral part of a computational process that attempts to emulate human perception. We present an algorithm which implements both modal and amodal completion and integrates a fully automatic decision-making mechanism for selecting between them. It proceeds directly from the outputs of the feature extraction module, infers descriptions in terms of overlapping layers, and labels junctions as T, L, X, and Y. We illustrate the approach on several challenging inputs, producing interpretations consistent with those of the human visual system.

CHAPTER 2

TENSOR VOTING

The tensor voting framework is an approach for perceptual organization that is able to infer salient structures based on the support the tokens which comprise them receive from their neighbors. It is based on the Gestalt principles of proximity and good continuation and can tolerate very large numbers of outliers. Data tokens are represented by tensors and their saliency is computed based on information propagated among neighboring tokens via tensor voting. The tokens can be any type of local primitive, such as points or surfels, that is localized in space and potentially has orientation preferences associated with it, but no dimensions. The framework has been developed over the past several years beginning in the work of Guy [24] which was followed by the work of Lee [49] and Tang [103]. Parts of the work presented here can also be found in [64].

A shortcoming of the original framework was its inability to detect terminations of the inferred perceptual structures. This has been addressed with the addition of first order information to the framework [112]. To avoid confusion we make the distinction between first and second order information throughout, even though the description of the first additions comes a later in Chapter 6. We begin by briefly going over other perceptual organization approaches and proceed to describe the original, second order formulation of tensor voting in 2-D and 3-D.

2.1 RELATED WORK

Perceptual organization has been an active research area since the beginning of the previous century based on the work of the Gestalt psychologists [42, 43, 118]. Important issues include noise robustness, initialization requirements, handling of discontinuities, flexibility in the types that can be represented, and computational complexity. This section reviews related work which can be classified in the following categories. More detailed descriptions can be found in [60, 64] where work on perceptual organization based on regularization, relaxation labeling, level set methods, clustering and robust estimation is also presented.

- symbolic methods
- methods based on local interactions
- methods inspired by psychophysiology and neuroscience.

Symbolic Methods. Following the paradigm set by Marr [58], many researchers developed methods for hierarchical grouping of symbolic data. Lowe [56] developed a system for 3-D object recognition based on perceptual organization of image edgels. Groupings are selected among the numerous possibilities according to the Gestalt principles, viewpoint invariance and low likelihood of being accidental formations. Later, Mohan and Nevatia [63] and Dolan and Riseman [17] also proposed perceptual organization approaches based on the Gestalt principles. Both are symbolic and operate in a hierarchical bottom-up fashion starting from edgels and increasing the level of abstraction at each iteration. The latter approach aims at inferring curvilinear structures, while the former aims at segmentation and extraction of 3-D scene descriptions from collations of features that have high likelihood of being projections of scene objects. Along the same lines is Jacobs' [34] technique for inferring salient convex groups among clutter since they most likely correspond to world objects. The criteria to determine the non-accidentalness of the potential structures are convexity, proximity and the contrast of the edgels.

Methods Based on Local Interactions. Shashua and Ullman [99] first addressed the issue of structural saliency and how prominent curves are formed from tokens that are not salient in isolation. They define a locally connected network that assigns a saliency value to every image location according to the length and smoothness of curvature of curves going through that location. In [79], Parent and Zucker infer trace points and their curvature based on spatial integration of local information. An important aspect of this method is its robustness to noise. This work was extended to surface inference in three dimensions by Sander and Zucker [86]. Sarkar and Boyer [89] employ a voting scheme to detect a hierarchy of tokens. Voting in parameter space has to be performed separately for each type of structure, thus making the computational complexity prohibitive for generalization to 3-D. The inability of previous techniques to simultaneously handle surfaces, curves and junctions was addressed in the precursor of our research, the work of Guy and Medioni [25, 26]. A unified framework where all types of perceptual structures can be represented is proposed along with a preliminary version of the voting scheme presented here. The major advantages of [25, 26] are noise robustness and computational efficiency, since it is not iterative. How this methodology evolved is presented in the remaining sections of this chapter.

Methods Inspired by Psychophysiology and Neuroscience. Finally, there is an important class of perceptual organization methods that are inspired by human perception and research in psychophysiology and neuroscience. Grossberg and Mingolla [22] and Grossberg and Todorovic [23] developed the *Boundary Contour System* and the *Feature Contour System* that can group fragmented and even illusory edges to form closed boundaries and regions by feature cooperation in a neural network. Heitger and von der Heydt [29], in a classic paper on neural contour processing, claim that elementary curves are grouped into contours via convolution with a set of orientation-selective kernels, whose responses decay with distance and difference in orientation. Williams and Jacobs [119] introduce the *stochastic completion fields* for contour grouping. Their probabilistic theory models the contour from a source to a sink as the motion of a particle performing a random walk. Particles decay after every step, thus minimizing the likelihood of completions that are not supported by the data or between distant points. Li [53] presents a contour integration model based on excitatory and inhibitory cells and a top-down feedback loop. What is more relevant to our research, that focuses on the pre-attentive, bottom-up process of perceptual grouping, is that connection strength decreases with distance, and that zero or low curvature alternatives are preferred to high curvature ones. The model for contour extraction of Yen and Finkel [123] is based on psychophysical and physiological evidence that has many similarities to ours. It employs a voting mechanism where votes, whose strength decays as a Gaussian function of distance, are cast along the tangent of the osculating circle. An excellent review of perceptual grouping techniques based on cooperation and inhibition fields can be found in [71]. Even though we do not attempt to present a biologically plausible system, the similarities between our framework and the ones presented in this paragraph are nevertheless encouraging.

Comparison With Our Approach.   Our methodology offers numerous advantages over previous work. Most other methods require oriented inputs to proceed. Using our method inputs can be oriented, unoriented or a combination of both. Our model-free approach allows us to handle arbitrary perceptual structures that adhere to Marr's "matter is cohesive" principle [58] only, and do not require predefined models that restrict the admissible solutions. Our representation is symbolic in the sense defined in [91]. This brings about advantages that include the ability to attach attributes to each token, and a greater flexibility in assigning meaningful interpretations to tokens. An important feature of our approach is that we are able to infer all possible types of perceptual structures, such as: volumes, surfaces, curves and junctions in 3-D simultaneously. This is possible without having to specify the type of structure we are interested in. Instead, analysis of the results of voting indicates the most likely type of structure at each position along with its normal and tangent orientations without having to specify in advance the desired type. To our knowledge, the tensor voting framework is the only methodology capable of this. Our voting function has many similarities with other voting-based methods, such as decay with distance and curvature [29, 53, 123], and the use of constant curvature paths [79, 89, 92, 123] that result in an eight-shaped voting field (in 2-D) [29, 123]. The major difference is that in our case, the votes cast are tensors and not scalars, therefore they are a lot richer in information. Each tensor simultaneously encodes *all* structure types allowing for a least commitment strategy until all information for a decision has been accumulated.

Furthermore, our results degrade much more gracefully in the presence of noise (see for example [25] and [60]).

2.2   TENSOR VOTING IN 2-D

This section introduces the tensor voting framework in 2-D. It begins by describing the original second order representation and voting of Medioni *et al.* [60]. It has been augmented with first order properties as part of this research, which is presented in detail in Chapter 6. To avoid confusion we will refer to the representation and voting of this chapter as *second order*, even though their first order counterparts have not been introduced yet. In the following sections we demonstrate how oriented and unoriented inputs can be encoded, and how they propagate their information to their neighbors in the form of votes. The orientation and magnitude of a second order vote cast from a unit oriented voter are chosen as in [25]. Based on the orientation and magnitude of this vote, the orientation and magnitude of the vote cast by an unoriented token can be derived. The appropriate information for all possible votes is contained in the the *stick* and *ball* voting fields. Finally, the present the way perceptual structures are inferred after analysis of the accumulated votes.

2.2.1   Second Order Representation in 2-D

The second order representation is in the form of a second order, symmetric, non-negative definite tensor which essentially indicates the saliency of each type of perceptual structure (curve, junction or region in 2-D) the token may belong to and its preferred normal and tangent orientations. Tokens cast second order votes to their neighbors according to the tensors they are associated with. A second order, symmetric, non-negative definite tensor is equivalent to a 2×2 matrix, or an ellipse. The axes of the ellipse are the eigenvectors of the tensor and their aspect ratio is the ratio of the eigenvalues. The major axis is the preferred *normal* orientation of a potential curve going through the location. The *shape* of the ellipse indicates the certainty of the preferred orientation. That is, an elongated ellipse represents a token with high certainty of orientation. Even further, a degenerate ellipse with only one non-zero eigenvalue represents a perfectly oriented point (a curvel). On the other hand, an ellipse with two equal eigenvalues represents a token with no preference for any orientation (Fig. 2.1(a)). The tensor's *size* encodes the saliency of the information encoded. Larger tensors convey more salient information than smaller ones. An arbitrary second order, symmetric, non-negative definite tensor can be decomposed as in the following equation:

$$T = \lambda_1 \hat{e}_1 \hat{e}_1^T + \lambda_2 \hat{e}_2 \hat{e}_2^T = (\lambda_1 - \lambda_2)\hat{e}_1 \hat{e}_1^T + \lambda_2(\hat{e}_1 \hat{e}_1^T + \hat{e}_2 \hat{e}_2^T) \qquad (2.1)$$

where λ_i are the eigenvalues in decreasing order and $\hat{e}_i$ are the corresponding eigenvectors (see also Fig. 2.1(b)). Note that the eigenvalues are non-negative since the tensor is non-negative

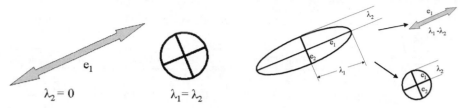

(a) Geometric illustration of saliency   (b) 2-D tensor decomposition

FIGURE 2.1: Illustration of 2-D second order symmetric tensors and decomposition of a tensor into its *stick* and *ball* components

definite. The first term in Eq. 2.1 corresponds to a degenerate elongated ellipsoid, termed hereafter the *stick tensor*, that indicates an elementary curve token with $\hat{e}_1$ as its curve *normal*. The second term corresponds to a circular disk, termed the *ball tensor*, that corresponds to a perceptual structure which has no preference of orientation or to a location where multiple orientations coexist. The size of the tensor indicates the certainty of the information represented by it. For instance, the size of the stick component $(\lambda_1 - \lambda_2)$ indicates curve saliency.

Based on the above, an elementary curve with normal $\vec{n}$ is represented by a stick tensor parallel to the normal, while an unoriented token is represented by a ball tensor. Note that curves are represented by their normals and not their tangents, for reasons that become apparent in higher dimensions. See Table 2.1 for how oriented and unoriented inputs are encoded and the equivalent ellipsoids and quadratic forms.

2.2.2   Second Order Voting in 2-D

After the inputs, oriented or unoriented, have been encoded with tensors, we examine how the information they contain is propagated to their neighbors. Given a token at O with normal $\vec{N}$ and a token at P that belong to the same smooth perceptual structure, the vote the token at O (the *voter*) casts at P (the *receiver*) has the orientation the receiver would have, if both the voter and receiver belonged to the same perceptual structure. The magnitude of the vote is a function of the confidence we have that the voter and receiver indeed belong to the same perceptual structure.

We first examine the case of a voter associated with a *stick tensor* and show how all other cases can be derived from it. We claim that, in the absence of other information, the arc of the *osculating circle* (the circle that shares the same normal as a curve at the given point) at O that goes through P is the most likely smooth path, since it maintains constant curvature. The center of the circle is denoted by C in Fig. 2.2. In case of straight continuation from O to P, the osculating circle degenerates to a straight line. Similar use of primitive circular arcs can also be found in [79, 89, 92, 123].

TABLE 2.1: Encoding oriented and unoriented 2-D inputs as 2-D second-order symmetric tensors

INPUT	SECOND ORDER TENSOR	EIGENVALUES	QUADRATIC FORM
oriented	stick tensor	$\lambda_1 = 1, \ \lambda_2 = 0$	$\begin{bmatrix} n_1^2 & n_1 n_2 \\ n_1 n_2 & n_2^2 \end{bmatrix}$
unoriented	ball tensor	$\lambda_1 = \lambda_2 = 1$	$\begin{bmatrix} 1 & 0 \\ 0 & 1 \end{bmatrix}$

As shown in Fig. 2.2, the second order vote is also a stick tensor and has a normal lying along the radius of the osculating circle at P. What remains to be defined is the magnitude of the vote. According to the Gestalt principles it should be a function of proximity and smooth continuation. The influence from one token to another should attenuate with distance, to minimize interference from unrelated tokens. The influence from one token to another should also attenuate curvature, to favor straight continuation over curved alternatives when both exist. Moreover, no vote is cast if the receiver is at an angle larger than $45°$ with respect to the tangent of the osculating circle at the voter. Similar restrictions on the fields appear also in [29, 53, 123]. The *saliency decay function* has the following form:

$$DF(s, \kappa, \sigma) = e^{-\left(\frac{s^2 + c\kappa^2}{\sigma^2}\right)} \tag{2.2}$$

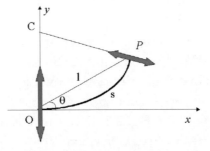

FIGURE 2.2: Second order vote cast by a stick tensor located at the origin

where s is the arc length OP, κ is the curvature, c controls the degree of decay with curvature, and σ is the *scale of voting*, which determines the effective neighborhood size. The parameter c is a function of the scale and is optimized to make the extension of two orthogonal line segments to from a right angle equally likely to the completion of the contour with a rounded corner [25]. Its value is given by:

$$c = \frac{-16\log(0.1) \times (\sigma - 1)}{\pi^2}. \tag{2.3}$$

Scale essentially controls the range within which tokens can influence other tokens. It can also be viewed as a measure of smoothness. A large scale favors long range interactions and enforces a higher degree of smoothness, aiding noise removal. A small scale makes the voting process more local, thus preserving details. Note that σ is the only free parameter in the system.

The 2-D second order stick vote for a unit stick voter located at the origin and aligned with the y-axis can be defined as follows as a function of the distance l between the voter and receiver and the angle θ, which is the angle between the tangent of the osculating circle at the voter and the line going through the voter and receiver (see Fig. 2.2).

$$\mathbf{S}_{SO}(l, \theta, \sigma) = DF(s, \kappa, \sigma) \begin{bmatrix} -\sin(2\theta) \\ \cos(2\theta) \end{bmatrix} \begin{bmatrix} -\sin(2\theta) & \cos(2\theta) \end{bmatrix}$$

$$s = \frac{\theta l}{\sin(\theta)}, \qquad \kappa = \frac{2\sin(\theta)}{l}. \tag{2.4}$$

The votes are also stick tensors. For stick tensors of arbitrary size the magnitude of the vote is given by Eq. 2.2 multiplied by the the size of the stick $\lambda_1 - \lambda_2$.

The ball tensor, which is the second elementary type of tensor in 2-D, has no preference of orientation, but still can cast meaningful information to other locations. The presence of two proximate unoriented tokens, the voter and the receiver, indicates a potential curve going through the two tokens. Votes cast by ball voters allow us to infer preferred orientations from unoriented tokens, thus minimizing initialization requirements. For simplicity we introduce the notation $\mathbf{B}_{so}(P)$ for the tensor which is the vote cast by a unit ball tensor at the origin to the receiver P. The derivation of the ball voting field $\mathbf{B}_{so}(P)$ from the stick voting field can be visualized as follows: the vote at P from a unit ball tensor at the origin O is the integration of the votes of stick tensors that span the space of all possible orientations. In 2-D, this is equivalent to a rotating stick tensor that spans the unit circle at O. The 2-D ball vote can be derived as a function of stick vote generation, according to:

$$\mathbf{B}_{so}(P) = \int_0^{2\pi} R_\theta^{-1} \mathbf{S}_{so}(R_\theta P) R_\theta^{-T} d\theta \tag{2.5}$$

where R_θ is the rotation matrix to align $\mathbf{S}$ with $\hat{e}_1$, the eigenvector corresponding to the maximum eigenvalue (the stick component), of the rotating tensor at P. In practice, the integration is approximated by tensor addition:

$$V = \sum_{i=1}^{K} \vec{v}_i \vec{v}_i^T \tag{2.6}$$

where V is the accumulated vote and $\vec{v}_i$ are the stick votes, in vector form. This is equivalent since a stick tensor has only one non-zero eigenvalue and can be expressed as the outer product of its only significant eigenvector. The stick votes from O to P cast by K stick tensors at angle intervals of $2\pi/K$ span the unit circle. Normalization has to be performed in order to make the energy emitted by a unit ball equal to that of a unit stick. The sum of the maximum eigenvalues of each vote is used as the measure of energy. As a result of the integration, the second order ball field does not contain purely stick or purely ball tensors, but arbitrary second order symmetric tensors. The field is radially symmetric, as expected, since the voter has no preferred orientation.

The voting process is identical whether the receiver contains a token or not, but we use the term **sparse voting** to describe a pass of voting where votes are cast to locations that contain tokens only, and the term **dense voting** for a pass of voting from the tokens to all locations within the neighborhood regardless of the presence of tokens. Receivers accumulate the votes cast to them by tensor addition.

2.2.3 Voting Fields

An interpretation of tensor voting can be made using the notion of *voting fields*, which can be thought of as emitting each token's preferred orientation to its neighborhood. The saliency values at a location is space are the combined effects of all voting fields that reach that particular location. Before the N-D extension of the tensor voting framework of [64], tensor voting was implemented using tensor fields to hold the votes. Votes from both stick and ball voters cast at receivers at various distances and angles were precomputed and stored in *voting fields*. These serve as look-up tables from which votes were retrieved by bilinear interpolation and could significantly speed up the voting process. Voting fields are briefly described here since they provide a useful illustration for the voting process.

The *fundamental voting field*, for which all fields can be derived, is the 2-D, second order, stick voting field. It contains at every position a tensor that is the vote cast there by a unit stick tensor located at the origin and aligned with the y axis. The shape of the field in 2-D can be seen in the upper part of Fig. 2.3(a). Depicted at every position is the eigenvector corresponding to the largest eigenvalue of the second order tensor contained there. Its size is proportional to the magnitude of the vote. To compute a vote cast by an arbitrary stick tensor, we need to align the

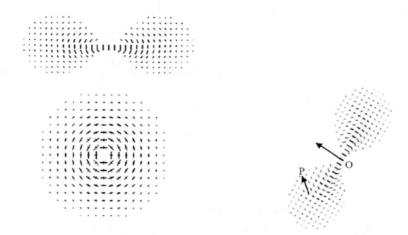

(a) The 2D stick and ball fields   (b) Stick vote cast from O to P

FIGURE 2.3: Voting fields in 2-D and alignment of the stick field with the data for vote generation

field with the orientation of the voter. Then we multiply the saliency of the vote that coincides with the receiver by the saliency of the arbitrary stick tensor, as in Fig. 2.3(b).

The *ball voting field* can be seen in the lower part of Fig. 2.3(a). The ball tensor has no preference of orientation, but still can cast meaningful information to other locations. The presence of two proximate unoriented tokens, the voter and the receiver, indicates a potential curve going through the two tokens. The ball voting field allows us to infer preferred orientations from unoriented tokens, thus minimizing initialization requirements. It is radially symmetric, as expected, since the voter has no preferred orientation.

Voting takes place in a finite neighborhood within which the magnitude of the votes cast remains significant. For example, we can find the maximum distance s_{max} from the voter at which the vote cast will have 1% of the voter's saliency, as follows:

$$e^{-\left(\frac{s_{max}^2}{\sigma^2}\right)} = 0.01 \qquad (2.7)$$

The size of this neighborhood is obviously a function of the scale σ. As described in section 2.2.1, any tensor can be decomposed into the basis components (stick and ball in 2-D) according to its eigensystem. Then, the corresponding fields can be aligned with each component. Votes are retrieved by simple look-up operations, and their magnitude is multiplied by the corresponding saliency. The votes cast by the stick component are multiplied by $\lambda_1 - \lambda_2$ and those of the ball component by λ_2.

2.2.4   Vote Analysis

Votes are cast from token to token and accumulated by tensor addition. Analysis of the second order votes can be performed once the eigensystem of the accumulated second order 2×2 tensor has been computed. Then the tensor can be decomposed into the stick and ball components:

$$T = (\lambda_1 - \lambda_2)\hat{e}_1\hat{e}_1^T + \lambda_2 \left(\hat{e}_1\hat{e}_1^T + \hat{e}_2\hat{e}_2^T \right) \tag{2.8}$$

where $\hat{e}_1\hat{e}_1^T$ is a *stick tensor*, and $\hat{e}_1\hat{e}_1^T + \hat{e}_2\hat{e}_2^T$ is a *ball tensor*. The following cases have to be considered:

- If $\lambda_1 - \lambda_2 > \lambda_2$, the saliency of the stick component is larger than that of the ball component and this indicates certainty of one normal orientation, therefore the token most likely belongs on a curve whose estimated normal is $\hat{e}_1$.

- If $\lambda_1 \approx \lambda_2 > 0$, the dominant component is the ball and there is no preference of orientation. This can occur either because all orientations are equally likely or because multiple orientations coexist at the location. This indicates either a token that belongs to a region, which is surrounded by neighbors from the same regions at all directions, or a junction where two or more curves intersect and multiple curve orientations are present simultaneously (see Fig. 2.4). Junctions can be discriminated from region inliers since their saliency is a distinct peak of λ_2. The saliency of region inliers is more evenly distributed.

- Finally, outliers receive only inconsistent, contradictory votes, so both eigenvalues are small.

2.2.5   Results in 2-D

An experiment on synthetic data can be seen in Fig. 2.5. The input is a set of points which are encoded as ball tensors before voting. After analysis of the eigensystem of the resulting tensors,

(a) Junction input   (b) Ball saliency map   (c) Region input   (d) Ball saliency map

FIGURE 2.4: Ball saliency maps at regions and junctions. Darker pixels in the saliency map correspond to higher saliency than lighter ones. The latter are characterized by a sharp peak of ball saliency.

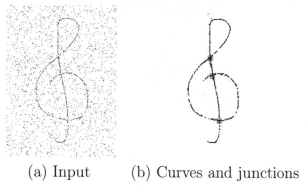

(a) Input   (b) Curves and junctions

FIGURE 2.5: Curves and junctions from a noisy point set. Junctions have been enlarged and marked as squares.

we can infer the most salient curve inliers and junctions. At the same time, we can remove the outliers due to their low saliency.

2.2.6   Quantitative Evaluation of Saliency Estimation

To evaluate the effectiveness of tensor voting in estimating the saliency of each input, we tested it with the datasets proposed in [120]. Each dataset consists of a foreground object represented by a sparse set of edgels super-imposed on a background texture, which is also represented as a sparse set of edgels. There are a total of nine foreground objects (fruit and vegetable contours), which are uniformly rescaled to fit 32×32 bounding boxes. There are also nine background textures which are rescaled to 64×64. The nine fruits and vegetables are: avocado, banana, lemon, peach, pear, red onion, sweet potato, tamarillo and yellow apple. The textures are taken from the MIT Media Lab texture database (http://vismod.media.mit.edu/vismod/ imagery/VisionTexture/vistex.html) and are: bark, brick, fabric, leaves, sand, stone, terrain, water and wood.

The goal is to detect the edgels of the foreground object, which align to form the largest salient contour. The background edgels come from an image of texture and are, therefore, less structured and do not produce alignments more salient than the foreground. The desired output is the N_f most salient edgels, where N_f is the number of edgels of the foreground. If they all belong to the foreground then performance is considered perfect. The reported error rates are the percentage of background edgels included in the N_f most salient. The difficulty comes from increasing the number of background edgels in each experiment. The SNR is defined as the ratio of foreground to background edgels in each dataset. Five SNRs ranging from 25% to 5% are used for each of the 81 combinations. All edgels are encoded as stick tensors of unit strength oriented at the given angles. After sparse voting, the given orientations are corrected and a

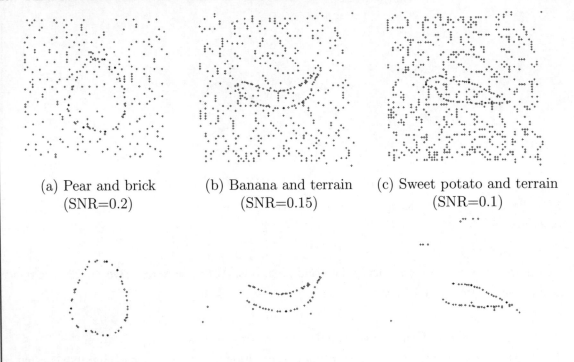

(a) Pear and brick   (b) Banana and terrain   (c) Sweet potato and terrain
(SNR=0.2)   (SNR=0.15)   (SNR=0.1)

(d) Output (FPR=5.88%)   (e) Output (FPR=16.39%)   (f) Output (FPR=24.19%)

FIGURE 2.6: Most salient inputs and false positive rates in typical examples from [120] at various SNRs.

second round of voting is performed. Since the error metric is based on the input positions, we only consider input locations in the second pass of voting. Figure 2.6 contains a few input and output pairs.

The false positive rates obtained by our algorithm (for $\sigma = 40$) can be seen in Table 2.2. It outperforms all the methods in [120], even though we do not consider closure, which plays a significant role in this experiment. The results we obtain are encouraging in our ongoing attempt to infer semantic descriptions from real images, even though phenomena such as junctions and occlusion have to be ignored, since the fruit appear transparent when encoded as sets edgels from their outlines in the input.

2.3 TENSOR VOTING IN 3-D

We proceed to the generalization of the framework in 3-D. No significant modifications need to be made, apart from taking into account that more types of perceptual structure exist in 3-D than in 2-D. In fact, the 2-D framework is a subset of the 3-D framework, which in

TABLE 2.2: False positive rates (FPR) for different signal to noise ratios for the data of [120]

SNR	FPR (%)
25	10.04
20	12.36
15	18.39
10	35.81
5	64.28

turn is a subset of the general N-D framework. The second order tensors are now 3-D, but vote generation can be easily derived from 2-D second order stick vote generation. In 3-D, the types of perceptual structure that have to be represented are regions (which are now volumes), surfaces, curves and junctions. The inputs can be either unoriented or oriented, in which case there are two types: elementary surfaces (*surfels*) or elementary curves (*curvels*).

2.3.1 Representation in 3-D

The representation of a token consists of a 3-D, second order, symmetric, non-negative definite tensor that encodes *saliency* as before. It is equivalent to a 3×3 matrix and a 3-D ellipsoid. The eigenvectors of the tensor are the axes of the ellipsoid and the corresponding eigenvalues are their lengths. The tensor can be decomposed as in the following equation:

$$
\begin{aligned}
T &= \lambda_1 \hat{e}_1 \hat{e}_1^T + \lambda_2 \hat{e}_2 \hat{e}_2^T + \lambda_3 \hat{e}_3 \hat{e}_3^T \\
&= (\lambda_1 - \lambda_2) \hat{e}_1 \hat{e}_1^T + (\lambda_2 - \lambda_3) \left(\hat{e}_1 \hat{e}_1^T + \hat{e}_2 \hat{e}_2^T \right) + \lambda_3 \left(\hat{e}_1 \hat{e}_1^T + \hat{e}_2 \hat{e}_2^T + \hat{e}_3 \hat{e}_3^T \right)
\end{aligned} \qquad (2.9)
$$

where λ_i are the eigenvalues in decreasing order and $\hat{e}_i$ are the corresponding eigenvectors (see also Fig. 2.7). The first term in Eq. 2.9 corresponds to a *3-D stick tensor*, that indicates an elementary surface token with $\hat{e}_1$ as its surface normal. The second term corresponds to a degenerate disk-shaped ellipsoid, termed hereafter the *plate tensor*, that indicates a curve or a surface intersection with $\hat{e}_3$ as its tangent, or, equivalently with $\hat{e}_1$ and $\hat{e}_2$ spanning the plane normal to the curve. Finally, the third term corresponds to a *3-D ball tensor*, that corresponds to a structure which has no preference of orientation. Table 2.3 shows how oriented and unoriented inputs are encoded and the equivalent ellipsoids and quadratic forms.

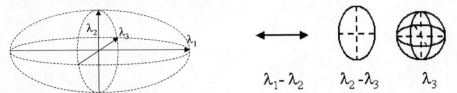

(a) A 3-D generic tensor (λ_i are its eigenvalues in descending order) (b) Decomposition into the *stick*, *plate* and *ball* components

FIGURE 2.7: A second order generic tensor and its decomposition in 3-D

TABLE 2.3: Encoding oriented and unoriented 2-D inputs as 2-D second order symmetric tensors

INPUT	TENSOR	EIGENVALUES	QUADRATIC FORM
surfel	stick tensor	$\lambda_1 = 1, \lambda_2 = \lambda_3 = 0$	$\begin{bmatrix} n_1^2 & n_1 n_2 & n_1 n_3 \\ n_1 n_2 & n_2^2 & n_2 n_3 \\ n_1 n_3 & n_2 n_3 & n_3^2 \end{bmatrix}$
curvel	plate tensor	$\lambda_1 = \lambda_2 = 1, \lambda_3 = 0$	$\mathbf{P}$ (see below)
unoriented	ball tensor	$\lambda_1 = \lambda_2 = \lambda_3 = 1$	$\begin{bmatrix} 1 & 0 & 0 \\ 0 & 1 & 0 \\ 0 & 0 & 1 \end{bmatrix}$

$$\mathbf{P} = \begin{bmatrix} n_{11}^2 + n_{21}^2 & n_{11}n_{12} + n_{21}n_{22} & n_{11}n_{13} + n_{21}n_{23} \\ n_{11}n_{12} + n_{21}n_{22} & n_{12}^2 + n_{22}^2 & n_{12}n_{13} + n_{22}n_{23} \\ n_{11}n_{13} + n_{21}n_{23} & n_{12}n_{13} + n_{22}n_{23} & n_{13}^2 + n_{23}^2 \end{bmatrix}$$

The representation using normals instead of tangents can be justified more easily in 3-D, where surfaces are arguably the most frequent type of structure. In 2-D, normal or tangent representations are equivalent. A surface patch in 3-D is represented by a stick tensor parallel to the patch's normal. A curve, which can also be viewed as a surface intersection, is represented by a plate tensor that is normal to the curve. All orientations orthogonal to the curve belong in the 2-D subspace defined by the plate tensor. Any two of these orientations that are orthogonal to each other can be used to initialize the plate tensor (see also Table 2.3). Adopting this representation allows a structure with $N - 1$ degrees of freedom in N-D (a curve in 2-D, a surface in 3-D) to be represented by a single orientation, while a tangent representation would require the definition of $N - 1$ vectors that form a basis for an (N-1)-D subspace. Assuming that this is the most frequent structure in the N-D space, our choice of representation makes vote generation for the stick tensor, which corresponds to the elementary (N-1)-D variety, the basis from which all other votes are derived. In addition, this choice makes the handling of intersections considerably easier. Using a representation based on normals, intersections are represented as the union of the normal spaces of each of the intersecting structures, which can be computed with the Gramm-Schmidt algorithm. On the other hand, using a representation based on tangents, the same operation would require the more cumbersome computation of the intersection of the tangent spaces.

2.3.2   Voting in 3-D

Identically to the 2-D case, voting begins with a set of oriented and unoriented tokens. We begin by showing how a voter with a purely stick tensor generates and casts votes, and then, derive the voting fields for the plate and ball cases. We chose to keep voting a function of only the position of the receiver relative to the voter and of the voter's preference of orientation. Therefore, we again address the problem of finding the smoothest path between the voter and receiver by fitting arcs of the osculating circle, as described in Section 2.2.1.

Note that the voter, the receiver and the stick tensor at the voter define a plane. The voting procedure is restricted on this plane, thus making it identical to the 2-D case. The second order vote, which is the surface normal at the receiver under the assumption that the voter and receiver belong to the same smooth surface, is also a purely stick tensor on the plane (see also Fig. 2.2). The magnitude of the vote is defined by the same *saliency decay function*, duplicated here for completeness:

$$DF(s, \kappa, \sigma) = e^{-(\frac{s^2 + c\kappa^2}{\sigma^2})} \qquad (2.10)$$

Sparse, token to token voting is performed to estimate the preferred orientation of tokens, followed by dense voting during which saliency is computed at every grid position.

From the perspective of voting fields, the 3-D stick voting field can be derived from the fundamental 2-D stick field by rotation about the voting stick, which is the axis of symmetry of the 3-D field. The visualization of the 2-D second order stick field in Fig. 2.3(a) is also a cut of the 3-D field that contains the stick tensor at the origin.

To show the derivation of a second order ball vote $\mathbf{B}_{so}(P)$ at P from a unit ball tensor at the origin O, we can visualize it as the integration of the votes of stick tensors that span the space of all possible orientations. In 2-D, this is equivalent to a rotating stick tensor that spans the unit circle at O, while in 3-D the stick tensor spans the unit sphere. The 3-D ball vote can be derived from the 3-D stick vote generation $\mathbf{S}_{so}(P)$, as follows:

$$\mathbf{B}(P)_{so} = \int_0^{2\pi} \int_0^{2\pi} R_{\theta\phi\psi}^{-1} \mathbf{S}_{so}(R_{\theta\phi\psi} P) R_{\theta\phi\psi}^{-T} \, d\phi d\psi \, |_{\theta=0} \qquad (2.11)$$

where $R_{\theta\phi\psi}$ is the rotation matrix to align $\mathbf{S}$ with $\hat{e}_1$, the eigenvector corresponding to the maximum eigenvalue (the stick component), of the rotating tensor at P, and θ, ϕ, ψ are rotation angles about the x, y, z axis respectively. The integration is approximated by tensor addition, $T = \sum \vec{v}_i \vec{v}_i^T$. Note that normalization has to be performed in order to make the energy emitted by a unit ball equal to that of a unit stick. The resulting voting field is radially symmetric, as expected, since the voter has no preferred orientation.

To complete the description of vote generation for the 3-D case, we need to describe the second order plate vote generation, denoted by $\mathbf{P}_{so}(P)$. Since the plate tensor encodes uncertainty of orientation around one axis, it can be derived by integrating the votes of a rotating stick tensor that spans the unit circle, in other words the plate tensor. The formal derivation is analogous to that of the ball voting fields and can be written as follows:

$$\mathbf{P}_{so}(P) = \int_0^{2\pi} R_{\theta\phi\psi}^{-1} \mathbf{S}_{so}(R_{\theta\phi\psi} P) R_{\theta\phi\psi}^{-T} \, d\psi \, |_{\theta=\phi=0} \qquad (2.12)$$

where θ, ϕ, ψ, and $R_{\theta\phi\psi}$ have the same meaning as in the previous equation. Normalization has to be performed in order to make the total energy of the ball and plate voting fields equal to that of the stick voting fields. The sum of the maximum eigenvalues of each vote is used as the measure of energy.

Voting by any 3-D tensor takes place by decomposing the tensor into its three components: the stick, the plate and the ball. Votes are retrieved from the appropriate voting field by look-up operations and are multiplied by the saliency of each component. Stick votes are weighted by $\lambda_1 - \lambda_2$, plate votes by $\lambda_2 - \lambda_3$ and ball votes by λ_3.

2.3.3 Vote Analysis

Analysis of the second order votes can be performed once the eigensystem of the accumulated second order 3×3 tensor has been computed. Then the tensor can be decomposed into the stick, plate and ball components:

$$T = (\lambda_1 - \lambda_2)\hat{e}_1\hat{e}_1^T + (\lambda_2 - \lambda_3)\left(\hat{e}_1\hat{e}_1^T + \hat{e}_2\hat{e}_2^T\right) + \lambda_3\left(\hat{e}_1\hat{e}_1^T + \hat{e}_2\hat{e}_2^T + \hat{e}_3\hat{e}_3^T\right). \qquad (2.13)$$

where $\hat{e}_1\hat{e}_1^T$ is a *stick tensor*, $\hat{e}_1\hat{e}_1^T + \hat{e}_2\hat{e}_2^T$ is a *plate tensor* and $\hat{e}_1\hat{e}_1^T + \hat{e}_2\hat{e}_2^T + \hat{e}_3\hat{e}_3^T$ is a *ball tensor*. The following cases have to be considered:

- If $\lambda_1 - \lambda_2 > \lambda_2 - \lambda_3$ and $\lambda_1 - \lambda_2 > \lambda_3$, the stick component is dominant. Thus the token most likely belongs on a surface whose normal is $\hat{e}_1$.

- If $\lambda_2 - \lambda_3 > \lambda_1 - \lambda_2$ and $\lambda_2 - \lambda_3 > \lambda_3$, the plate component is dominant. In this case the token belongs on a curve or a surface intersection. The normal plane to the curve or the surface orientation is spanned by $\hat{e}_1$ and $\hat{e}_2$. Equivalently, $\hat{e}_3$ is the tangent.

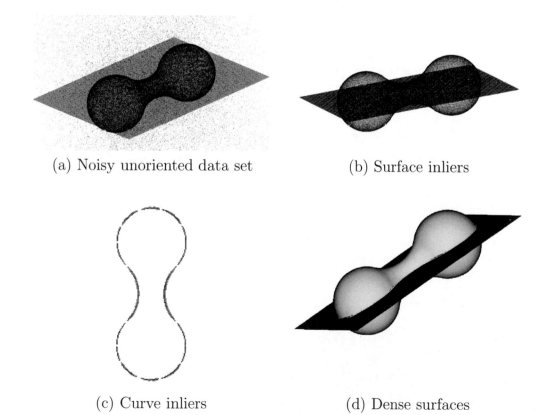

(a) Noisy unoriented data set   (b) Surface inliers

(c) Curve inliers   (d) Dense surfaces

FIGURE 2.8: Inference of surfaces and surface intersections from noisy data

- If $\lambda_3 > \lambda_1 - \lambda_2$ and $\lambda_3 > \lambda_2 - \lambda_3$, the ball component is dominant and the token has no preference of orientation. It is either a junction or it belongs in a volume. Junctions can be discriminated from volume inliers since they are distinct peaks of λ_3.

- Outliers receive only inconsistent, contradictory votes, so both eigenvalues are small.

2.3.4   Results in 3-D

Due to space considerations and to the fact that more challenging experiments are presented in Chapters 3 and 6 we present results on just just one synthetic 3-D dataset. The example in Fig. 2.8 illustrates the simultaneous inference of surfaces and curves. The input consists of a "peanut" and a plane, encoded as unoriented points contaminated by random uniformly distributed noise (Fig. 2.8(a)). The "peanut" is empty inside, except for the presence of noise, which has an equal probability of being anywhere in space. Figure 2.8(b) shows the detected surface inliers, after tokens with low saliency have been removed. Figure 2.8(c) shows the curve inliers, that is the tokens that lie at the intersection of the two surfaces. Finally, Fig. 2.8(d) shows the extracted dense surfaces.

CHAPTER 3

Stereo Vision from a Perceptual Organization Perspective

In this chapter, we address the fundamental problem of matching pixels in two static images. Significant progress has been made in this area, but the correspondence problem has not been completely solved due mostly to occlusion and lack of texture. We propose an approach that addresses these difficulties within a perceptual organization framework, taking into account both binocular and monocular sources of information. Initially, matching candidates for all pixels are generated by a combination of several matching techniques. The matching candidates are then reconstructed in disparity space. In this space, perceptual organization takes place in 3D neighborhoods and, thus, does not suffer from problems associated with scanline or image neighborhoods. The assumption is that correct matches form salient coherent surfaces, while wrong matching candidates do not align. Surface saliency, therefore, is used as the criterion to disambiguate matches. The matching candidates that are kept are grouped into smooth layers. Surface overextensions, which are systematic errors due to occlusion, can be corrected at this stage by ensuring that each match is consistent in color with its neighbors of the same layer in *both* images. Matches that are not consistent in both images are most likely errors due to the foreground overextending and covering occluded parts of the image. These are removed and the labeled surfaces are refined. Finally, the projections of the refined surfaces on both images can be used to obtain disparity hypotheses for pixels that remain unmatched. The final disparities are selected after a second tensor voting stage, during which information is propagated from more reliable pixels to less reliable ones. The proposed framework takes into account both geometric and photometric smoothness.

3.1 INTRODUCTION

The premise of shape from stereo comes from the fact that, in a set of two or more images of a static scene, world points appear on the images at different disparities depending on their distance from the cameras. Establishing pixel correspondences on real images, though, is far from trivial. Projective and photometric distortion, sensor noise, occlusion, lack of texture, and repetitive patterns make matching the most difficult stage of a stereo algorithm. Here we focus on occlusion and insufficient or ambiguous texture, which are inherent difficulties of the depicted scene, and not of the sensors.

To address these problems, we propose a stereo algorithm that operates as a perceptual organization process in the 3D disparity space, keeping in mind that false matches will most likely occur in textureless areas, and close to depth discontinuities. Since binocular processing has limitations in these areas, we use monocular information to overcome them. We begin by generating matching hypotheses for every pixel within a flexible framework that allows the use of matches generated by any matching technique reported in the literature. These matches are reconstructed in a 3D (x, y, d) space, where d denotes the disparity. In this space, the correct matches align to form surfaces, while the wrong ones do not form salient structures. We can infer a set of reliable matches based on the support they receive from their neighbors as surface inliers via tensor voting. These reliable matches are grouped into layers. Note that the term layer is used interchangeably with surface, since by layer we indicate a smooth, but not necessarily planar, surface in 3D disparity space. The surfaces are refined by rejecting matches that are consistent in color with their neighbors in both images. The refined, segmented surfaces serve as the "unambiguous component" as defined in [88] to guide disparity estimation for the remaining pixels.

Segmentation using geometric properties is arguably the most significant contribution of this research effort. It provides very rich information on the position, orientation, and appearance of the surfaces in the scene. Moreover, grouping in 3D circumvents many of the difficulties associated with image segmentation. It is also a process that treats both images symmetrically, unlike other approaches where only one of the two images is segmented. Candidate disparities for unmatched pixels are generated after examining the color similarity of each unmatched pixel with its nearby layers. If the color of the pixel is compatible with the color distribution of a nearby layer, disparity hypotheses are generated based on the existing layer disparities and the disparity gradient limit constraint [81]. Tensor voting is then performed locally and votes are collected at the hypothesized locations. Only matches from the appropriate layer cast votes to each candidate match. The hypothesis that is the smoothest continuation of the surface is kept as the disparity for the pixel under consideration. In addition, assuming that the occluded surfaces are partially visible and that the occluded parts are smooth continuations of the visible ones, we are able to extrapolate them and estimate the depth of monocularly visible pixels. Under this scheme, smoothness with respect to both shape, in the form of surface continuity, and appearance, in the form of color similarity, is taken into account before disparities are assigned to unmatched pixels.

This chapter is organized as follows: related work is reviewed in the next section; Section 3.3 is an overview of the algorithm; Section 3.4 describes the initial matching stage; Section 3.5 the detection of correct matches using tensor voting; Section 3.6 the segmentation and refinement process; Section 3.7 the disparity computation for unmatched pixels; Section 3.8 contains experimental results; Section 3.9 summarizes our approach to stereo; Section 3.10 briefly presents other computer vision research in 3D using tensor voting.

3.2   RELATED WORK

In this section, we review research on stereo related to ours. We focus on area-based and pixel-based methods since their goal is a dense disparity map. Feature-based approaches are not covered, even though the matches they produce can be integrated into our framework. We also consider only approaches that handle discontinuities and occlusions. The input images are assumed to be rectified and the epipolar lines to coincide with the scanlines. If this is not the case, the images can be rectified using methods such as [128].

The problem of stereo is often decomposed as the establishment of pixel correspondences and surface reconstruction, in Euclidean or disparity space. These two processes, however, are strongly linked, since the reconstructed pixel correspondences form the scene surfaces, while at the same time, the positions of the surfaces dictate pixel correspondences in the images. In the remainder of this chapter, we describe how surface saliency is used as the criterion for the correctness of matches, as in [50, 51]. Arguably, the first approach where surface reconstruction does not follow but interacts with feature correspondence is that of Hoff and Ahuja [30]. They integrate matching and surface interpolation to ensure surface smoothness, except at depth discontinuities and creases. Edge points are detected as features and matched across the two images at three resolutions. Planar and quadratic surface patches are successively fitted and possible depth or orientation discontinuities are detected at each resolution. The patches that fit the matched features best are selected while the interpolated surfaces determine the disparities of unmatched pixels.

Research on dense area-based stereo with explicit treatment of occlusion includes numerous approaches (see [12, 96] for comprehensive reviews of stereo algorithms). They can be categorized as follows: local, global, and approaches with extended local support, such as the one we propose. Local methods attempt to solve the correspondence problem using local operators in relatively small neighborhoods. Local methods using adaptive windows were proposed by Kanade and Okutomi [38] and Veksler [113]. Birchfield and Tomasi [7] introduced a new pixel dissimilarity measure that alleviates the effects of sampling, which are a major source of errors when one attempts to establish pixel correspondence. Their experiments, as those of [102] and ours, demonstrate the usefulness of this measure, which we use in the work presented here.

On the other hand, global methods arrive at disparity assignments by optimizing a global cost function that usually includes penalties for pixel dissimilarity and violation of the smoothness constraint. The latter introduces a bias for constant disparity at neighboring pixels, thus favoring frontoparallel planes. Chronologically, the first global optimization approaches to stereo were based on dynamic programming. Since dynamic programming addresses the problem as a set of 1D subproblems on each epipolar line separately, these approaches suffer from inconsistencies between adjacent epipolar lines that appear as streaking artifacts. Efforts to address this weakness were published as early as 1985, when Ohta and Kanade used edges to provide intrascanline constraints [77]. Recent work also attempts to mitigate streaking by enforcing interscanline constraints, but the problem is not entirely eliminated. Dynamic programming methods that explicitly model occlusion include [4, 5, 7, 9, 20, 31].

Consistency among epipolar lines is guaranteed by using graph cuts to optimize the objective function, since they operate in 2D. Roy and Cox [83] find the disparity surface as the minimum cut of an undirected graph. In this framework, scanlines are no longer optimized independently, with interscanline coherence enforced later in a heuristic way, but smoothness is enforced globally over the entire image. Other stereo approaches based on graph cuts include [32, 44, 45].

Between these two extremes are approaches that are neither "winner-take-all" at the local level, nor global. They rely on more reliable matches to estimate the disparities of less reliable ones. Following Marr and Poggio [59], Zitnick and Kanade [129] employed the support and inhibition mechanism of cooperative stereo to ensure the propagation of correct disparities and the uniqueness of matches with respect to both images without having to rely on the ordering constraint. Reliable matches without competitors are used to reinforce matches that are compatible with them, while at the same time, they eliminate those that contradict them, progressively disambiguating more pixels. A cooperative approach using deterministic relaxation and explicit visibility handling was proposed by Luo and Burkhardt [57]. Zhang and Kambhamettu [125] extend the cooperative framework from single pixels to image regions.

A different method of aggregating support is nonlinear diffusion, proposed by Scharstein and Szeliski [95], where disparity estimates are propagated to neighboring points in disparity space until convergence. Sun et al. [100, 101] formulate the problem as an MRF with explicit handling of occlusions. In the belief propagation framework, information is passed to adjacent pixels in the form of messages whose weight also takes into account image segmentation. The process is iterative and has similar properties with nonlinear diffusion.

Sara [88] formally defines and computes the largest unambiguous component of stereo matching, which can be used as a basis for the estimation of more unreliable disparities. Other similar approaches include those of Szeliski and Scharstein [102] and Zhang and Shan [126] who start from the most reliable matches and allow the most certain disparities to guide the estimation of less certain ones, while occlusions are explicitly labeled.

The final class of methods reviewed here utilizes monocular color cues (image segmentation) to guide disparity estimation. Birchfield and Tomasi [8] cast the problem of correspondence as image segmentation followed by the estimation of affine transformations between the

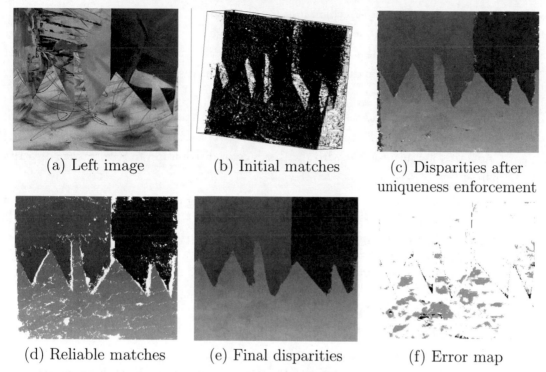

(a) Left image   (b) Initial matches   (c) Disparities after
                                                  uniqueness enforcement

(d) Reliable matches   (e) Final disparities   (f) Error map

FIGURE 3.1: Overview of the processing steps for the "Sawtooth" dataset. The initial matches have been rotated so that the multiple candidates for each pixel are visible. Black pixels in the error map indicate errors greater than 1 disparity level, gray pixels correspond to errors between 0.5 and 1 disparity level, while white pixels are correct (or occluded and thus ignored).

images for each segment that can account for slanted surfaces. Lin and Tomasi [54] propose a framework where 3D shape is estimated by fitting splines, while 2D support is based on image segmentation. Processing alternates between these two steps until convergence.

3.3   OVERVIEW OF OUR APPROACH

Our approach [70] to the derivation of dense disparity maps from rectified image pairs falls into the category of area-based stereo since we attempt to infer matches for every pixel using matching windows. It has four steps, which are illustrated in Fig. 3.1, for the "Sawtooth" stereo pair. The steps are as follows:

- *Initial matching*, where matching hypotheses are generated for every pixel by a combination of different matching techniques. The dataset now includes multiple candidate matches for each pixel and can be seen in Fig. 3.1(b).

- *Detection of correct matches*, which uses tensor voting to infer the correct matches from the unorganized point cloud of the previous stage as inliers of salient surfaces. After tensor voting, uniqueness is enforced with respect to surface saliency and the dataset contains at most one candidate match per pixel. The disparity map can be seen in Fig. 3.1(c).

- *Surface grouping and refinement*, during which the matches are grouped into smooth surfaces, using the estimated surface orientations. These surfaces are refined by removing points that are inconsistent with the layer's color distribution to result in the disparity map of Fig. 3.1(d).

- *Disparity estimation for unmatched pixels*, where the goal is to assign disparities that ensure smoothness in terms of both surface orientation and color properties of the layers. The final disparity map and the error map can be seen in Figs. 3.1(e) and (f).

These steps are presented along with experimental evaluations in Sections 3.4 through 3.7.

3.4   INITIAL MATCHING

A large number of matching techniques have been proposed in the literature [96]. They have different strengths and weaknesses and each is more suitable for certain types of pixels. We propose a scheme for combining a variety of matching techniques, thus taking advantage of their combined strengths. For the results presented in this chapter, four matching techniques are used, but any type of matching operator can be integrated into the framework. The techniques used here are as follows:

- A small (typically 5×5) normalized cross correlation window, which is small enough to capture details and only assumes constant disparity over small windows of the image. This technique is referred to as the "correlation window" in the remainder of this chapter. The correlation coefficients for all possible disparities values of each pixel are computed and we keep all peaks of the correlation function, with magnitudes that are comparable to the maximum for the pixel, since they are good candidates for correct pixel correspondences. They are used as inputs for the tensor voting stage, where the decisions are made based on surface saliency. The correlation coefficients is not used since it can be affected by factors such as repetitive patterns or the degree of texture of one surface over the other.

- A *shiftable* normalized cross correlation window of the same size as the above, which achieves good performance near discontinuities. It is referred to as the "shiftable window" in the remainder of this chapter. The limitation of window-based matching is that, no matter how small the window is, pixels from two or more surfaces are included in it at discontinuities. By not centering the window on the pixel under consideration,

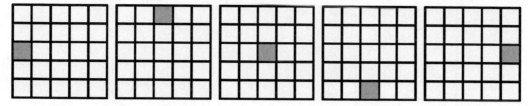

FIGURE 3.2: The five shiftable windows applied for each disparity choice at every pixel. The shaded square corresponds to the pixel under consideration. The same window is applied to the target image.

we can find a shift that includes as many pixels as possible from the same surface as the pixel under consideration. See Fig. 3.2 for the five windows used here. Given a pixel in the reference image, we compute cross correlation for each disparity level for five different window shifts around the pixel under consideration, and keep the one with the maximum correlation coefficient as the score for that disparity level. As with correlation windows, we keep all significant peaks of the score function as candidate matches.

- A 25 × 25 normalized cross correlation window, which is applied only at pixels where the standard deviation of the three color channels is less than 20. The use of such a big window over the entire image would be catastrophic, but it is effective when applied only in virtually textureless regions, where smaller windows completely fail to detect correct matches. This technique is referred to as the "large window".

- A *symmetric interval* matching window (typically 7 × 7) with truncated cost function as in [102]. This is referred to as the "interval window". The final matching technique is very different from the above, not only because we use the matching cost of [7], but mostly because of the truncation of the cost for each pixel at a certain level. That makes the behavior robust against pixels from different surfaces that have been included in the window. Both images are linearly interpolated, as in [102], along the x-axis so that samples exist in half-pixel intervals. The intensity of each pixel, in each of the three color channels, is now represented as the interval between the minimum and maximum value of the intensity at the integer pixel position and the half-pixel positions before and after it on the scanline, as shown in Fig. 3.3.

Numerically, the cost for matching pixel (x_L, y) in the left image with pixel (x_R, y) in the right image is the minimum distance between the two intervals, which is given by the following equation and is zero if they overlap:

$$C(x_L, x_R, y) = \sum_{c \in \{R, G, B\}} \min \Big\{ \text{dist}(I_{Lc}(x_i, y), I_{Rc}(x_j, y)), c_{\text{trunc}} : \qquad (3.1)$$

$$x_i \in \left[x_L - \frac{1}{2} \quad x_L + \frac{1}{2} \right], x_j \in \left[x_R - \frac{1}{2} \quad x_R + \frac{1}{2} \right] \Big\}. \qquad (3.2)$$

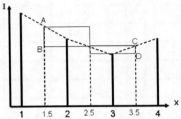 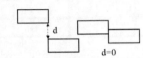

(a) Interpolation and interval representation   (b) Distance between intervals

FIGURE 3.3: Symmetric interval matching. Both images are interpolated to double the number of pixels per row. Then, each pixel is represented by the interval defined by its own intensity value and the values of its two neighbors. For instance, the pixel at position 2 is represented using its own value and the values of positions 1.5 and 2.5. These three values produce the interval labeled AB above. Pixel dissimilarity is defined as the distance between the two intervals in the left and the right image, and not between an interval and a pixel.

The summation is over the three color channels and dist() is the Euclidean distance between the value of a color channel I_{Lc} in the left image and I_{Rc} in the right image. If the distance for any channel exceeds the truncation parameter c_{trunc}, the total cost is set to $3c_{\text{trunc}}$. Typical values for c_{trunc} are between 3 and 10. Even though, statistically, the performance of interval windows is slightly worse than that of the shiftable windows, they are useful because they produce correct disparity estimates for pixels where the other windows fail. This is due to the different nature of the dissimilarity measure and the robust formulation we use.

Note that the typical window sizes are for image resolutions similar to those of the Middlebury image pairs, which range from 284×216 to 450×375. Larger window sizes would most likely be necessary for higher resolution images.

Each matching technique is repeated using the right image as reference and the left as target. This increases the true positive rate especially near discontinuities, where the presence of occluded pixels in the reference window affects the results of matching. When the other image is used as reference, these pixels do not appear in the reference window. A simple parabolic fit [96] is used for subpixel accuracy, mainly because it makes slanted or curved surfaces appear continuous and not staircase-like. In addition to the increased number of correct detections, the combination of these matching techniques offers the advantage that the failures of a particular technique are not detrimental to the success of the algorithm, as long as the majority of the operators do not produce the same erroneous disparities. Our experiments have also shown that the errors produced by small windows, such as 5×5 and 7×7 used here, are randomly spread in space and do not usually align to form nonexistent structures. This property is important for our methodology that is based on the perceptual organization, due to good alignment, of candidate matches in space. On the other hand, the large window, which is more susceptible to systematic errors, is never applied near discontinuities.

3.5 SELECTION OF MATCHES AS SURFACE INLIERS

Even if the local matching operators perform exceptionally well, the problem of stereo in its entirety, taking into account occlusions and discontinuities, cannot be fully solved at the pixel level. Support for each potential match has to be aggregated, so that the confidence of correct matches is increased and outliers are made explicit. Aggregation in 1D neighborhoods is only motivated by computational simplicity and its shortcomings are well documented. While graph cut based methods have achieved outstanding results, the choice of an appropriate energy function is not an easy task. Energy functions whose global minima can be found with current optimization techniques do not necessarily model the phenomenon of stereovision in its most general form. In most cases, the disparity assignment that achieves the globally minimal energy is not the ground truth disparity of the scene. This occurs because the energy function has to satisfy certain properties to be suitable for minimization using algorithms such as graph cuts or belief propagation [107]. For instance, the penalization of disparity changes between neighboring pixels makes these approaches well suited for scenes that consist of frontoparallel planes and prefers staircase-looking solutions for slanted or curved surfaces. Here, following the approach of Lee et al. [51] we aggregate support in 3D neighborhoods via tensor voting. Pixels that are close in one image but are projections of remote world points do not interact. Fig. 3.4 shows four points that project relatively close to each in the image. Points A and B, which are close in 3D, and therefore are likely to belong to the same scene surface, interact

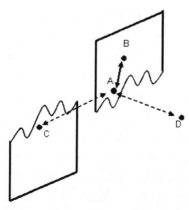

FIGURE 3.4: Voting in 3D neighborhoods eliminates interference between adjacent pixels from different layers.

strongly with each other. On the other hand, points A and C that are close in the image but not in 3D, and therefore are most likely projections of unrelated surfaces, do not vote to each other. Finally, point D, which is isolated in 3D and is probably generated by an error in the initial matching stage, receives no support as an inlier of a salient surface. After accumulating support by tensor voting, candidate matches that are consistent with their neighbors have high surface saliency, which validates them as correct matches.

The goal here is to address stereo as a perceptual organization problem, based on the premise that the correct matches should form coherent surfaces in disparity space. The input is a cloud of points in a 3D space $(x, y, zscale \times d)$, where $z\,scale$ is a constant used to make the input less flat with respect to the d-axis, since disparity space is usually a lot flatter than actual (x, y, z). Its typical value is 8 and the sensitivity is very low. The quantitative matching scores are disregarded and all candidate matches are initialized as unoriented ball tensors with saliency equal to 1. If two or more matches fall within the same $(x, y, zscale \times d)$ voxel their initial saliencies are added, thus increasing the confidence of candidate matches confirmed by multiple matching techniques. Since d is estimated with subpixel accuracy each integer disparity level has $zscale$ possible subpixel levels. Therefore, quantization occurs at a finer resolution and the dimensions of each voxel are $pixel_{size} \times pixel_{size} \times zscale$.

The inputs, encoded as ball tensors, cast votes to their neighbors. When voting is completed, the surface saliency of each candidate match can be computed as the difference between the two largest eigenvalues of the tensor. Regardless of the criterion used, certain constraints have traditionally been applied at this stage to disambiguate pixel correspondences. Since the ordering constraint is violated by scene configurations that are not unlikely, such as the presence of thin foreground objects, we do not enforce it. Its popularity in the literature is mostly as a byproduct of optimization techniques. As optimization techniques have improved, most researchers have abandoned the ordering constraint. The uniqueness constraint, which states that in the absence of transparency there should be at most one match for each pixel, should also be enforced carefully. As Ogale and Aloimonos [76] point out, if scene surfaces exhibit horizontal slant (that is, if the epipolar line in the image and the intersection of the epipolar plane and the scene surface are not parallel), then M pixels in one image necessarily correspond to N pixels in the other image. Therefore, the requirement for a strict one-to-one correspondence for all pixels results in labeling $|M - N|$ pixels as occluded. These pixels that are interleaved with matched pixels, however, are perfectly visible in both images, just not at integer coordinate positions. Keeping this observation in mind, we only enforce uniqueness as a postprocessing step allowing at most one match for each pixel of the reference image in order to derive a dense disparity map. More than one pixel of the reference image is allowed to correspond to the same pixel of the target image (with integer or subpixel disparities) if the surface appears wider in the reference image. Thus, visible pixels are not marked as being occluded due to surface slant. A similar approach has also been presented in [100].

In our approach uniqueness is enforced with respect to the left image by retaining the candidate with the highest surface saliency for every pixel. We do not enforce uniqueness with respect to the right image since it is violated by slanted surfaces which project to a different number of pixels on each image. Since the objective is disparity estimation for every pixel in the reference image, uniqueness applies to that image only. Surface saliency is a more reliable criterion for the selection of correct matches than the score of a local matching operator. This is because it requires that candidate matches, identified as such by local operators, also form coherent surfaces in 3D. This scheme is capable of rejecting false positive responses of the local operators, which is not possible at the local level. The resulting datasets still contain errors, which are corrected at the next stage. The most frequently used, besides smoothness, are the ordering and uniqueness constraints.

3.6   SURFACE GROUPING AND REFINEMENT

A rather safe conclusion that can be drawn from the Middlebury Stereo Evaluation (http://cat.middlebury.edu/stereo/) is that the use of monocular information, such as color, contributes to the performance of a stereo algorithm. In [68, 70], we proposed a novel way of integrating monocular information that requires very few assumptions about the scene and does not fail when image segmentation fails. Instead of trying to identify the scene surfaces by their projections on the images via their color properties, we try to infer them in disparity space via surface grouping. Candidate matches that were retained after tensor voting are grouped into smooth surfaces based on their 3D positions and estimated surface normals. Then these surfaces are reprojected to both images, and points that are inconsistent with the other points of the surface in terms of color distribution in either image are rejected. This step removes erroneous matches for occluded pixels, which are usually assigned with the disparity of the foreground. They are removed since they do not project to the same surface in both images, and thus the color distributions are inconsistent. Under this scheme, both images are treated symmetrically, unlike most segmentation-based methods where only the reference image is segmented. Furthermore, we do not attempt to segment the image, but instead solve a simpler problem: grouping points, which were selected as surface inliers, into smooth 3D surfaces.

Grouping candidate matches that have not been rejected in layers is achieved using a simple growing scheme. By layers here we mean surfaces with smooth variation of surface normal. They do not have to be planar, and the points that belong to them do not have to form one connected component. Labeling starts from seed matches that have maximum surface saliency. Since the input to this stage is rather dense and includes candidate matches for a large percentage of the matches, we only examine the eight nearest neighbors of the seed. If they

(a) Left image of "Venus" pair   (b) Ground truth

FIGURE 3.5: The left image of the "Venus" pair and the ground truth depth map. The left circle covers two surfaces, while the right one covers a single surface. Both regions show that pixels with different colors could belong to the same surface, but that pixels with similar colors are more likely to belong to the same surface. We use this observation to clean up the noisy initial surfaces from the grouping stage. A nonparametric representation turns out to be very effective despite its simplicity.

are smooth continuations of the growing surface they are added to it and their neighbors are also considered for addition. The disparity gradient limit constraint dictates that the maximum disparity jump between two pixels of the same surface is 1. When no more matching candidates can be added to the surface, the unlabeled point with maximum surface saliency is selected as the next seed. Small surfaces comprising less than 0.5% of image pixels are removed, since they are probably noisy patches, unless they are compatible with a larger nearby surface in terms of both position and orientation. Support from a larger surface means that the small part is most likely correct, but due to occlusion or failure of the matching operators is not connected to the main part of the surface. After this step, the dataset consists of a set of labeled surfaces, which contain errors mostly due to foreground overextension. A number of candidate matches that survived uniqueness enforcement while not being parts of large salient surfaces are also removed here. These include wrong matches in uniform areas, which are not aligned with the correct matches.

The next step is the refinement of the layers. The goal is to remove the overextensions of the foreground by ensuring that the color properties of the pixels, which are the projections of the grouped points, are *locally* consistent within each layer. Color consistency of a pixel is checked by computing the ratio of pixels of the same layer with similar color to the current pixel over the total number of pixels of the layer within the neighborhood. For example, the left red circle in Fig. 3.5(a) covers two different surfaces. One could try to segment them and be successful in this case, but we argue that by simply rejecting red pixels that have been assigned disparities not supported by other red pixels we can clean up the original noisy surfaces under the minimum number of assumptions. This nonparametric representation allows us to handle cases like the newspaper surface in the right red circle where a simple parametric model is not sufficient to describe a surface with interleaved black and white pixels. In this case, a mixture of two Gaussian distributions is required and more complex cases can be encountered very often in practice. Our approach does not need to model the color distribution of the surfaces, but only relies on the assumption that neighboring pixels with similar colors are more likely to have similar depths.

This color consistency check is repeated in the target image and if the current assignment does not correspond to the maximum ratio *in both images*, then the pixel is removed from the layer. The color similarity ratio for pixel (x_0, y_0) in the left image with layer i can be computed according to the following equation,

$$R_i(x_0, y_0) = \frac{\sum_{(x,y)\in N} T(\mathrm{lab}(x, y) = i \text{ AND dist}(I_L(x, y), I_l(x_0, y_0) < c_{\mathrm{thr}}))}{\sum_{(x,y)\in N} T(\mathrm{lab}(x, y) = i))}, \qquad (3.3)$$

where $T()$ is a test function that is 1 if its argument is true, $\mathrm{lab}()$ is the label of a pixel, and c_{thr} is a color distance threshold in RGB space, typically equal to the c_{trunc} parameter of the interval windows. The same is applied for the right image for pixel $(x_0 - d_0, y_0)$. The size of the neighborhood is the second and final parameter of this stage. It can be set equal to the range of the voting field during tensor voting.

This step corrects surface overextension that occurs near occlusions, since the overextensions are usually not color consistent in both images and are thus detected and removed. Table 3.1 shows the total number of candidate matches and errors before and after refinement for the

TABLE 3.1: Total and Wrong Matches in Each Dataset Before and After Surface Grouping and Refinement

IMAGE PAIR	TOTAL BEFORE	ERROR RATE BEFORE (%)	TOTAL AFTER	ERROR RATE AFTER (%)
Tsukuba	84810	5.31	69666	1.33
Sawtooth	144808	2.95	136894	1.08
Venus	147320	6.16	132480	1.24
Map	48657	0.44	45985	0.05
Cones	132856	4.27	126599	3.41
Teddy	135862	7.24	121951	4.97

four Middlebury image pairs. The disparity maps for the "Sawtooth" example before and after grouping and refinement can be seen in Figs. 3.1(c) and (d).

3.7   DISPARITY ESTIMATION FOR UNMATCHED PIXELS

The goal of this stage is to generate candidate matches for the remaining unmatched pixels. Given the already estimated disparities and labels for a large set of pixels, there is more information available now that can enhance our ability to estimate the missing disparities. We opt for a progressive approach, under which only the most reliable correspondences are allowed at first. These are correspondences that satisfy strict geometric and color requirements in both images. The requirements become less strict as we proceed.

Given an unmatched pixel in the reference image, we examine its neighborhood for layers to which the pixel can be assigned. Color similarity ratios are computed for the pixel with respect to all layers as in Eq. (3.3). The layer with the maximum ratio is selected as the potential layer for the pixel. Then, we need to generate a range of disparities for the pixel. This is done by examining the disparity values of the selected layer's pixels in the neighborhood. The range is extended according to the disparity gradient limit constraint, which holds perfectly in the case of rectified parallel stereo pairs. Disparity hypotheses are verified one by one on the target image by computing similarity ratios, unless they are occluded. In the latter case, we allow occluded surfaces to grow underneath the occluding ones. On the other hand, we do not allow new matches to occlude existing consistent matches. Votes are collected at valid potential matches in disparity space, as before, with the only difference being that only matches from the appropriate layer cast votes (see Fig. 3.6). The most salient among the potential matches is selected and added to the layer, since it is the one that ensures the smoothest surface continuation.

For the results presented here, we applied the following progressive growing scheme, which has two parameters: c_{thr}, the color threshold used for computing the similarity ratios, and σ_3, the scale of voting for densification which also defines the size of the neighborhood in which similarity ratios are computed. For the first iteration, we initialize the parameters with $c_{\text{thr}} = 1$ and $\sigma_3^2 = 20$. These are very strict requirements and have to be satisfied on both images for a disparity hypothesis to be valid. Votes are accumulated on the valid hypotheses that also do not occlude any existing matches, and the most salient continuation is selected. Then, we repeat the process without requiring consistency with the target image and add more matches, which are usually for occluded pixels that are similar to their unoccluded neighbors. The added matches are generally correct, but valid hypotheses cannot be generated for all pixels. In the second iteration we increment both c_{thr} and σ_3^2 by their initial values and repeat the process. The choice of parameters here is not critical. For instance, maintaining a constant σ_3 produces very similar results. For the experiments shown here, both parameters are increased by constant increments at each iteration until convergence.

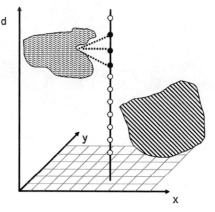

FIGURE 3.6: Candidate generation for unmatched pixels based on segmented layers. The unmatched pixel is compatible with the left surface only, thus votes are collected at disparity hypotheses generated by matches of the left surface (marked with black dots). Also note that only matches from the appropriate layer vote at each candidate.

Typically, there are a few pixels that cannot be resolved because they exhibit low similarity to all layers, or because they are specular or in shadows. Candidates for these pixels are generated based on the disparities of all neighboring pixels and votes are collected at the candidate locations in disparity space. Again, the most salient ones are selected. We opt to use surface smoothness at this stage instead of image correlation, or other image-based criteria, since we are dealing with pixels where the initial matching and color consistency failed to produce a consistent match.

3.8   EXPERIMENTAL RESULTS

This section contains results on the color versions of the four image pairs of [96] and the two proposed in [97], which are available online at http://cat.middlebury.edu/stereo/. *All six* examples were processed with identical parameters. The initial matching in all cases was done using the four matching operators presented in Section 3.4 using both the left and right image as reference. The correlation and shiftable windows were 5×5. The interval windows were 7×7 with the truncation parameter set at 5. The large window was 25×25, applied at pixels with intensity variance less than 20. For the large windows only, pixels with normalized score below 20% of the average were rejected. The scale of the voting field for the detection of correct matches was $\sigma^2 = 50$, which corresponds to a voting radius of 14, or a neighborhood of $29 \times 29 \times 29$. Refinement was performed with a voting radius of 18 and c_{thr} equal to 5. In the final stage, c_{thr} was initialized as 1 and incremented by 1 for 25 iterations, while σ_3^2 was initialized as 20 and incremented by 20 at each iteration.

TABLE 3.2: Error Rates for the Original Middlebury Image Pairs

IMAGE PAIR	UNOCC-LUDED (%)	RANK	DISCONTI-NUITIES (%)	RANK	TEXTURELESS (%)	RANK
Tsukuba	1.51	11	7.96	12	2.02	24
Sawtooth	0.70	12	4.35	11	0.50	26
Venus	1.09	12	13.95	26	1.39	16
Map	1.31	24	11.47	26	—	—

A second surface refinement operation was performed to remove errors around the surface discontinuities. This time, the voting radius was significantly smaller, set equal to 7, since we are only interested in correcting the borders of each surface. The value of c_{thr}, on the other hand, was equal to 40, to allow larger color variation within each surface. The parameters for the final stage were identical with those of the previous paragraph.

The error metric reported in the tables is that proposed in [96], where matches are considered erroneous if they correspond to unoccluded image pixels and their disparity error is greater than one integer disparity level. Table 3.2 contains the error rates we achieved, as well as the rank our algorithm would achieve among the 38 algorithms in the evaluation. The error rates reflect the number of errors larger than 1 disparity level for all unoccluded pixels, for pixels near discontinuities and for textureless pixels. We have rounded the disparities to integer values for this evaluation. We refer readers to the Middlebury Stereo Evaluation web page for results obtained by other methods. Based on the results for all unoccluded pixels, our algorithm would rank fifteenth in the evaluation as of July 5, 2005. As with all methods that take color explicitly into account, performance on the "Map" is not as good as that achieved by methods that do not use monocular information due to the random textures in the image.

Tables 3.3 and 3.4 report our results for the new version of the Middlebury Stereo Evaluation that includes "Tsukuba," "Venus," and the two image pairs introduced in [97]. The new image pairs contain curved and slanted surfaces, with different degrees of detail and texture, and are, thus, more challenging. This is more pronounced for methods that make the assumption that scene surfaces are planar and parallel to the image plane. This assumption is explicitly made when one penalizes disparity differences between neighboring pixels. Table 3.3 contains the error rates when the acceptable error is set to 1 disparity level, while Table 3.4 contains the error rates when the acceptable error is set only to 0.5 disparity level. This demonstrates the capability of the algorithms to estimate precise subpixel disparities. We have not rounded the disparities in this case. For the new evaluation, the error rate over all pixels,

TABLE 3.3: Quantitative Evaluation for the New Middlebury Image Pairs (acceptable error at 1.0 disparity level)

IMAGE PAIR	UNOCC- LUDED (%)	RANK	ALL (%)	RANK	DISCONTINUITIES (%)	RANK
Tsukuba	3.79	9	4.79	9	8.86	6
Venus	1.23	4	1.88	5	11.5	9
Teddy	9.76	5	17.0	5	24.0	8
Cones	4.38	3	11.4	4	12.2	5

including the occluded ones, has replaced the evaluation over textureless pixels. The ranks are among the 12 algorithms that are being evaluated, as of July 5, 2005. Considering performance at unoccluded pixels, our results are tied at the fourth place when the acceptable error is 1, and rank third when it is 0.5.

Figs. 3.7 and 3.8 show the final disparity map and the error map for the "Venus," "Tsukuba," "Map," "Cones," and "Teddy" image pairs. The results for "Sawtooth" appear in Fig. 3.1. White in the error maps indicates an error less than 0.5 disparity level or occluded pixel, gray indicates an error between 0.5 and 1 disparity level (acceptable) and black indicates large errors above 1 disparity level.

3.9   DISCUSSION

We have presented a novel stereo algorithm that addresses the limitations of binocular matching by incorporating monocular information. We use tensor voting to infer surface saliency and use it as a criterion for deciding on the correctness of matches as in [50, 51]. However, the

TABLE 3.4: Quantitative Evaluation for the New Middlebury Image Pairs (acceptable error at 0.5 disparity level)

IMAGE PAIR	UNOCC- LUDED (%)	RANK	ALL (%)	RANK	DISCONTINUITIES (%)	RANK
Tsukuba	25.5	11	26.2	11	21.2	8
Venus	3.32	1	4.12	1	14.6	2
Teddy	14.6	3	21.8	4	33.3	4
Cones	7.05	2	14.5	3	17.4	3

FIGURE 3.7: Left images: final disparity maps and error maps for the "Venus", "Tsukuba," and "Map" image pairs from the Middlebury Stereo Evaluation.

quality of the experimental results depends heavily on the inputs to the voting process. The superior performance of the new algorithm is due to the flexible initial matching stage and the combination of the geometric and photometric consistency we enforce on the surfaces. Textured pixels away from depth discontinuities can be easily resolved by even naive stereo algorithms. As stated in the introduction, we aimed at reducing the errors at untextured parts of the image and near depth discontinuities which cause occlusion. In our approach, the typical phenomenon of the overextension of foreground surfaces over occluded pixels is mitigated by removing from the dataset candidate matches that are not consistent with their neighboring pixels in both images. On the other hand, surface smoothness is the main factor that guides the matching of uniform pixels.

Arguably, the most significant contribution is the segmentation into layers based on geometric properties and not appearance. We claim that this is advantageous over other methods that use color-based segmentation, since it utilizes the already computed disparities which

FIGURE 3.8: Left images: final disparity maps and error maps for the "Cones" and "Teddy" image pairs from the Middlebury Stereo Evaluation.

are powerful cues for grouping. In fact, grouping candidate matches in 3D based on good continuation is a considerably easier problem than image segmentation. This scheme allows us to treat both images symmetrically and provides estimates for the layer color distribution even if it varies significantly throughout the layer. The choice of a local, nonparametric color representation allows us to handle surfaces with texture or heterogeneous and varying color distributions, such as those in the "Venus" images, on which image segmentation may be hard. This representation is used at the layer refinement stage to eliminate mostly foreground overextensions.

A second significant contribution is the initial matching stage that allows the integration of any matching technique without any modification to subsequent modules. The use of a large number of matching operators, applied to both images, can be viewed as another form of consensus. While all operators fail for certain pixels, the same failures are usually not repeated, with the same disparity values, by other operators. Our experiments show that the results of combining the four techniques we used over all the image pairs are superior to those generated by using a smaller set of them. Even though a particular matching technique may produce systematic errors for a particular image pair, its inclusion is beneficial when all six image pairs are considered.

Adhering to the principles set out in Chapter 1, we employ a least-commitment strategy and avoid the use of constraints that are violated by usual scene configurations. One such constraint is the requirement that adjacent pixels have the same disparity to avoid incurring some penalty. While this constraint aids the optimization process of many approaches, it becomes an approximation for scenes that do not consist of frontoparallel surfaces. Processing in 3D via tensor voting enforces the more general constraint of good continuation and eliminates interference between adjacent pixels from different world surfaces without having to assess penalties on them. In our work, the assumption that scene surfaces are frontoparallel is only made in the initial matching stage, when all pixels in a small window are assumed to have the same disparity. After this point, the surfaces are never assumed to be anything other than continuous. We also do not use the ordering constraint, which was introduced to facilitate dynamic programming. The uniqueness constraint is applied cautiously, since it is viewpoint dependent and its results do not hold for the target image.

Our algorithm fails when surfaces are entirely missed at the initial matching stage or when they are entirely removed at the layer refinement stage. We are not able to grow surfaces that are not included in the data before the final stage. On the other hand, we are able to smoothly extend partially visible surfaces to infer the disparities of occluded pixels, assuming that occluded surfaces do not abruptly change orientation.

3.10   OTHER 3D COMPUTER VISION RESEARCH

In this section, we briefly present two other research thrusts that have been developed within the tensor voting framework.

3.10.1 Multiple-View Stereo

An extension of our binocular work to multiple frames has been published in [65, 66]. Here we very briefly describe the contributions of the latter paper. As in the binocular case, the premise is that correct pixel correspondences align to form the scene surfaces which are more salient than any potential alignments of wrong matches. The main contribution is a computational framework for the inference of dense descriptions from multiple-view stereo with a far wider range of permissible camera configurations. Thus far, research on dense multiple-view stereo has evolved along three axes: computation of scene approximations in the form of visual hulls; merging of depth maps derived from simple configurations, such as binocular or trinocular; multiple-view stereo with restricted camera placement. These approaches are either suboptimal, since they do not maximize the use of available information, or cannot be applied to general camera configurations.

We present an approach that allows truly general camera placement, employs a world-centered representation and is able to process large numbers of potential pixel matches, typically in the order of a few millions, efficiently. No images are privileged and features are not required to appear in more than two views. The only restriction on camera placement is that cameras

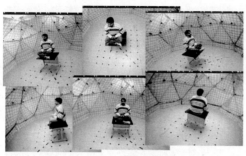 

(a) Some input images   (b) Ground truth

FIGURE 3.9: Some of the input images of the "meditation" set captured at the CMU dome (a few cameras are visible in each image) and a view of the reconstructed points.

must be placed in pairs in such a way that for each camera there exists at least another one with a similar viewpoint that allows for automatic correlation-based dense pixel matching. One could place such camera pairs arbitrarily in space with no other considerations for image overlap or relationships between the locations of camera centers and the scene. Moreover, unlike other leading multiple-view reconstruction methods [19, 45, 47, 124], we do not segment and discard the "background" but attempt to reconstruct it together with the foreground. Our approach does not involve binocular processing other than the detection of tentative pixel correspondences. The inference of scene surfaces is based on the premise that correct pixel correspondences, reconstructed in 3D, form salient, coherent surfaces, while wrong correspondences form less coherent structures. The tensor voting framework is suitable for this task since it can process the very large datasets we generate with reasonable computational complexity. In [66] we present results on challenging datasets captured for the Virtualized Reality project of the Robotics Institute of Carnegie Mellon University and distributed freely at http://www-2.cs.cmu.edu/virtualized-reality. Fig. 3.9 shows a few of the input images and a view of the reconstructed points from the "meditation" dataset.

3.10.2 Tracking

A different computer vision application in a 3D space is tracking. An approach that can track objects in motion when observed by a fixed camera, with severe occlusions, merging and splitting objects and defects in the detection was presented in [46]. We first detect regions corresponding to moving objects in each frame, which become the tokens, then try to establish their trajectory. The method is based on implementing spatiotemporal continuity in a $2D + t$ space, which represents the position of the moving regions in the images through time. A key difference

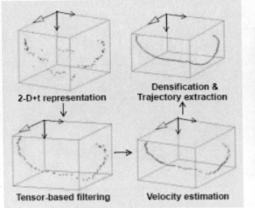

(a) Processing steps   (b) Tracking results

FIGURE 3.10: Illustration of the processing steps from short fragmented trajectories to long reliable tracks and some results on real videos with multiple moving people.

with regular tensor voting, due to the presence of the time axis, is the fact that voting has to be oriented toward the time direction. By this we mean that a voting token at time t does not vote for points in the same frame, but only supports good continuations in previous and subsequent frames. The steps of the process are highlighted in Fig. 3.10(a) and some results are shown in Fig. 3.10(b). This work was extended to multiple cameras in [39].

CHAPTER 4

Tensor Voting in ND

In this chapter, we describe a major contribution to the tensor voting framework. Even though this work is very recent, we feel that tensor voting in high-dimensional spaces may turn out to be among the most potent algorithms for manifold learning and related tasks. Initial results on a variety of machine learning tasks can be seen in Chapter 5. The motivation for this work came from the observation that many problems, from a broad range of scientific domains, can be addressed within the tensor voting framework. In most cases, addressing these problems requires efficient, local, data-driven algorithms with high noise robustness. The Gestalt principles of proximity, similarity, and good continuation, which have been identified as the factors that make configurations in two and three dimensions salient, still apply in spaces of higher dimensionality. Therefore, the tensor voting framework, which combines many of the desired properties, seems to be a well-suited approach. The main limitation that prevents its wide application to problems in high dimensions is the exponential increase in computational and storage demands as the dimensionality of the space grows.

4.1 INTRODUCTION

The tensor voting framework, in its preliminary version [25], is an attempt at the implementation of two Gestalt principles, namely proximity and good continuation, for grouping generic tokens in 2D. The 2D domain has always been the main focus of research in perceptual organization, beginning with the research of Köhler [43], Wertheimer [118], and Koffka [42] up to the recent work that was reviewed in Section 2.1. The generalization to 3D is straightforward, since salient groupings can be detected by the human visual system based on the same principles. Guy and Medioni extended the framework to 3D in [26]. Other perceptual organization approaches with 3D implementations include [78, 86]. Their number is considerably smaller than that of 2D methodologies, mostly due to the exponential increase in computational complexity with the dimensionality of the space. Regardless of the computational feasibility of an implementation, the same grouping principles apply to spaces with even higher dimensions. For instance, Tang et al. [106] observed that pixel correspondences can be viewed as points in the 8D space of free parameters of the fundamental matrix. Correct correspondences align to form a hyperplane in that space, while wrong correspondences are randomly distributed. By applying tensor voting in 8D, the authors were able to infer the dominant hyperplane and the desired parameters of the fundamental matrix. This work was reformulated as a 4D problem and solved for the case of multiple independently moving objects by Tong et al. [111].

Even though the applicability of tensor voting as an unsupervised learning technique in high-dimensional spaces seems to have merit, a general implementation is not practical. This is mostly due to computational complexity and storage requirements in N dimensions. The storage requirements for each token are $O(N \times N)$, which is acceptable. It can be reduced with a sparse storage scheme, but that would increase the necessary computations to retrieve information. The bottleneck is the generation and storage of the voting fields, the number of which is equal to the dimensionality of the space. For instance, a second-order voting field in 10D with k samples per axis requires storage for 10^k $N \times N$ tensors, which need to be computed via numerical integration over 10 variables. Moreover, the likelihood of each precomputed vote being used decreases with the dimensionality. Thus, the use of precomputed voting fields soon becomes impractical as dimensionality grows. At the same time, the computation of votes "on the fly" as they become necessary is also computationally expensive. Here, we propose a simplified vote generation scheme that bypasses the computation of uncertainty and allows the generation of votes from arbitrary tensors in arbitrary dimensions with a computational cost that is linear with respect to the dimensionality of the space. Storage requirements are limited to storing the tensors at each token since voting fields are not used any more.

This chapter is organized as follows: the next section points out the limitations of the original implementation of tensor voting; Section 4.3 describes the new voting scheme including a comparison with the original implementation; Section 4.4 offers a comparison of the new implementation with the original one of [60]; Section 4.5 shows results on various high-dimensional computer vision problems; Section 4.6 summarizes the contributions of the chapter.

4.2 LIMITATIONS OF ORIGINAL IMPLEMENTATION

The major limitations of the original implementation of tensor voting that render its generalization to ND impractical are related to the use of precomputed voting fields. The difficulties are due to the lack of a closed-form solution for the integrals required to compute votes cast by tensors that are not purely sticks. For example, the computation of a vote cast by a plate tensor in 3D, as shown in Eq. 2.11, requires the integration of the votes cast by a rotating stick tensor over two angles that span 360°. Approximating the computation by a summation with 1° steps results in 129,600 stick vote computations. In general, the computation of a vote cast by a tensor with N normals requires $N - 1$ nested summations to span the normal space and 360^{N-1} stick vote computations. An additional problem is caused by the fact that this simple sampling scheme of constant angular steps does not produce a uniform sampling in polar coordinates if more than one summation is necessary. More uniform samplings are possible, especially if one aligns the poles of the hypersphere with the nonvoting direction from the voter to receiver. A uniform sampling improves the accuracy, but does not reduce computational complexity.

In terms of storage, one can exploit symmetries, such as that the stick field is essentially 2D, regardless of the dimensionality of the space, and that the ball field is 1D, since it is a function of distance only. Nevertheless, the storage of the voting fields is impossible even for relatively modest values of N. A smaller weakness of precomputed voting fields is the fact that votes at nongrid positions are not exact, since they are produced by linear interpolation between grid positions.

Another limitation of the previous implementation is the use of a multiple level priority queue for data storage. The weaknesses of this data structure is that search operations are optimal if the axes have been assigned optimal priority. Otherwise, search may degenerate to a $O(M)$ operation, where M is the number of tokens. This becomes more pronounced as dimensionality increases.

4.3   TENSOR VOTING IN HIGH-DIMENSIONAL SPACES

In this section, we describe in detail the new vote generation scheme. We adopt the data representation and vote analysis of [60] and the previous chapters, but use the ANN k-d tree of [1] as the data structure that stores the data and retrieves the neighbors. Data representation and vote analysis are also presented in this chapter for completeness, as well as to illustrate the differences between 2D, 3D, and high-dimensional spaces. Note that efficiency with this formulation is considerably higher than our initial attempt in [69].

4.3.1   Data Representation

The representation of a point is still a second-order, symmetric, nonnegative definite tensor. It can represent the structure of a manifold going through the point by encoding the normals to the manifold as eigenvectors of the tensor that correspond to nonzero eigenvalues. The tangents are represented as eigenvectors that correspond to zero eigenvalues. A point in an ND hyperplane has one normal and $N - 1$ tangents, and thus is represented by a tensor with one nonzero eigenvalue associated with an eigenvector parallel to the hyperplane's normal. The remaining $N - 1$ eigenvalues are zero. A point in a 2D manifold in ND has two tangents and $N - 2$ normals, and thus is represented by a tensor with two zero eigenvalues associated with eigenvectors that span the tangent space of the manifold. The tensor also has $N - 2$ nonzero eigenvalues (typically set to 1) whose corresponding eigenvectors span the manifold's normal space.

The tensors can be formed by the summation of the direct products ($\vec{n}\vec{n}^T$) of the eigenvectors that span the normal space of the manifold. The tensor at a point on a manifold of dimensionality d, with $\vec{n}_i$ spanning the normal space, can be computed as follows:

$$T = \sum_{i=1}^{d} \vec{n}_i \vec{n}_i^T. \tag{4.1}$$

As before, a ball tensor is viewed as one having all possible normals and is encoded as an $N \times N$ identity matrix. Any point on a manifold of known dimensionality and orientation can be encoded in this representation by appropriately constructed tensors, as in Eq. (4.1).

On the other hand, given an ND second-order, symmetric, nonnegative definite tensor, the type of structure encoded in it can be found by examining its eigensystem. Any tensor that has these properties can be decomposed as in the following equation,

$$\mathbf{T} = \sum_{d=1}^{N} \lambda_d \hat{e}_d \hat{e}_d^T \tag{4.2}$$

$$= (\lambda_1 - \lambda_2)\hat{e}_1\hat{e}_1^T + (\lambda_2 - \lambda_3)(\hat{e}_1\hat{e}_1^T + \hat{e}_2\hat{e}_2^T) + \cdots + \lambda_N(\hat{e}_1\hat{e}_1^T + \hat{e}_2\hat{e}_2^T + ... + \hat{e}_N\hat{e}_N^T) \tag{4.3}$$

$$= \sum_{d=1}^{N-1} \left[(\lambda_d - \lambda_{d+1}) \sum_{k=1}^{d} \hat{e}_d\hat{e}_d^T \right] + \lambda_N(\hat{e}_1\hat{e}_1^T + \cdots + \hat{e}_N\hat{e}_N^T) \tag{4.4}$$

where λ_d are the eigenvalues in descending order and $\hat{e}_d$ are the corresponding eigenvectors. The tensor simultaneously encodes *all* possible types of structures. The saliency of the type that has d normals is encoded in the difference $\lambda_d - \lambda_{d+1}$. If a hard decision on the dimensionality is required, we assign the point to the type with the maximum confidence.

4.3.2 The Voting Process

In this section, we describe a novel vote generation scheme that does not require integration. As in the original formulation, the eigenstructure of the vote represents the normal and tangent spaces that the receiver would have, if the voter and receiver belong to the same smooth structure. What is missing is the uncertainty in each vote that resulted from the accumulation of the votes cast by the rotating stick tensors during the computation of the voting fields. The new votes are cast directly from the voter to the receiver and are not retrieved from precomputed voting fields. They have perfect certainty in the information they convey. The uncertainty now comes only from the accumulation of votes from different votes at each token.

We begin by examining the case of a voter that is associated with a stick tensor of unit length, which is identical with the original formulation since it can be directly computed. The

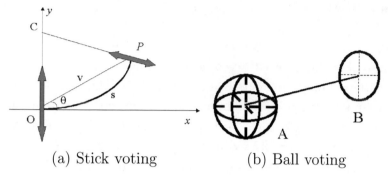

(a) Stick voting   (b) Ball voting

FIGURE 4.1: Vote generation for a stick and a ball voter. The votes are functions of the position of voter A and receiver B and the tensor of the voter.

vote is generated according to the following equation:

$$S(s, \theta) = e^{-\left(\frac{s^2 + c\kappa^2}{\sigma^2}\right)} \begin{bmatrix} -\sin(2\theta) \\ \cos(2\theta) \end{bmatrix} [-\sin(2\theta) \; \cos(2\theta)] \qquad (4.5)$$

$$s = \frac{\theta \|\vec{v}\|}{\sin(\theta)}, \qquad \kappa = \frac{2\sin(\theta)}{\|\vec{v}\|}. \qquad (4.6)$$

As in the 2D and 3D s is the length of the arc between the voter and receiver, and κ is its curvature (see Fig. 2.2(a)), σ is the scale of voting, and c is a constant. No vote is generated if angle θ is greater than 45°. Also, the field is truncated to the extent where the magnitude of the vote is greater than 3% of the magnitude of the voter. The vote as defined above is on the plane defined by A, B and the normal at A. Regardless of the dimensionality of the space, stick vote generation always takes place in a 2D subspace defined by the position of the voter and the receiver and the orientation of the voter. Thus, this operation is identical in any space between

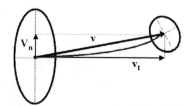

FIGURE 4.2: Vote generation for generic tensors. The voter here is a tensor with two normals in 3D. The vector connecting the voter and receiver is decomposed into $\vec{v}_n$ and $\vec{v}_t$ that lie in the normal and tangent space of the voter. A new basis that includes $\vec{v}_n$ is defined for the normal space and each basis component casts a stick vote. Only the vote generated by the orientation parallel to $\vec{v}_n$ is not parallel to the normal space. Tensor addition of the stick votes produces the combined vote.

2 and N dimensions. After the vote has been computed, it has to be transformed to the ND space.

Regarding the generation of ball votes, we propose the following direct computation. It is based on the observation that the vote generated by a ball voter propagates the voter's preference for a straight line that connects it to the receiver (Fig. 2.2(b)). The straight line is the simplest and smoothest continuation from a point to another point in the absence of other information. Thus, the vote generated by a ball is a tensor that spans the $(N-1)$D normal space of the line and has one zero eigenvalue associated with the eigenvector that is parallel to the line. Its magnitude is a function of the distance between the two points, since curvature is zero. Taking these observations into account, the ball vote can be constructed by subtracting the direct product of the tangent vector from a full rank tensor with equal eigenvalues (i.e., the identity matrix). The resulting tensor is attenuated by the same Gaussian weight according to the distance between the voter and the receiver,

$$\mathbf{T}(s,\theta) = e^{-\left(\frac{s^2}{\sigma^2}\right)} \left(\mathcal{I} - \frac{\vec{v}\vec{v}^T}{\|\vec{v}\vec{v}^T\|} \right). \tag{4.7}$$

where $\vec{v}$ is a unit vector parallel to the line connecting the voter and the receiver.

To complete the description of vote generation, we need to describe the case of a tensor that has d equal eigenvalues, where d is not equal to 1 or N. (The description applies to these cases too, but we use the above direct computations, which are faster.) Let $\vec{v}$ again be the vector connecting the voting and receiving points. It can be decomposed into $\vec{v}_t$ in the tangent space of the voter and $\vec{v}_n$ in the normal space. The new vote generation process is based on the observation that curvature in Eq. (4.5) is not a factor when θ is zero, or, in other words, if the voting stick is orthogonal to $\vec{v}_n$. We can exploit this by defining a new basis for the normal space of the voter that includes $\vec{v}_n$. The new basis is computed using the Gramm–Schmidt procedure. The vote is then constructed as the tensor addition of the votes cast by stick tensors parallel to the new basis vectors. Among those votes, only the one generated by the stick tensor parallel to $\vec{v}_n$ is not parallel to the normal space of the voter and curvature has to be considered. All other votes are a function of the length of $\vec{v}_t$ only. See Fig. 4.2 for an illustration in 3D. Analytically, the vote is computed as the summation of d stick votes cast by the new basis of the normal space. Let N_S denote the normal space of the voter and let $\vec{b}_i$, $i \in [1, d]$, be a basis for it with $\vec{b}_1$ being parallel to $\vec{v}_n$. If vote() is the function that, given a unit vector as an argument, generates the stick vote from a unit stick tensor parallel to the argument to the receiver, then the vote from a generic tensor with normal space N is given by

$$\mathbf{T} = \text{vote}(\vec{b}_1) + \sum_{i \in [2,d]} \text{vote}(\vec{b}_i). \tag{4.8}$$

In the above equation, all the terms are pure stick tensors parallel to the voters, except the first one which is affected by the curvature of the path connecting the voter and receiver and is orthogonal to it. Therefore, computation of the last $d - 1$ terms is equivalent to applying the Gaussian weight to the voting sticks and adding them at the position of the receiver. Only one vote requires a full computation of orientation and magnitude. This makes the proposed scheme computationally inexpensive.

Tensors with unequal eigenvalues are decomposed before voting according to Eq. (4.2). Then, each component votes separately and the vote is weighted by $\lambda_d - \lambda_{d+1}$, except the ball component whose vote is weighted by λ_D.

4.3.3 Vote Analysis

Vote analysis is a direct generalization of the original formulation, with the only difference being that $N + 1$ structure types are possible in an N D space. Each point casts a vote to all neighbors within the distance at which vote magnitude attenuates to 3% of the maximum. The votes are accumulated at each point by tensor addition. The eigensystem of the resulting tensor is computed and the tensor is decomposed as in Eq. (4.2). The estimate of local intrinsic dimensionality is given by the maximum gap in the eigenvalues. For instance, if $\lambda_1 - \lambda_2$ is the maximum difference between two successive eigenvalues, the dominant component of the tensor is that which has one normal. Quantitative results in dimensionality estimation are presented in the next chapter. In general, if the maximum eigenvalue spread is $\lambda_d - \lambda_{d+1}$, the estimated local intrinsic dimensionality is $N - d$, and the manifold has d normals and $N - d$ tangents. Moreover, the first d eigenvectors that correspond to the largest eigenvalues are the normals to the manifold, and the remaining eigenvectors are the tangents.

4.4 COMPARISON AGAINST THE OLD TENSOR VOTING IMPLEMENTATION

In this section, we evaluate the effectiveness of the new implementation by performing the same experiments in 3D using both old and new vote generation schemes. Fig. 4.3 shows a cut, containing the voter, of the 3D ball voting field computed using the old and the new implementation. Shown are the projections of the eigenvectors on the selected plane after voting by a ball voter placed on the plane.

We do not attempt to directly compare vote magnitudes, since they are defined differently. In the old implementation, the total energy of the ball field is normalized to be equal to that of the stick field. In the new implementation, the magnitude of the ball vote is the same as that of the stick vote. This makes the total energy of the new field higher than that of the stick field, since there is no attenuation due to curvature at any orientation, and voting takes place at all orientations, since the 45° cut-off only applies to stick voters.

(a) Old implementation   (b) New implementation

FIGURE 4.3: Visualization of the ball voting field using the old and the new implementation. Shown are the curve normals, as well as the tangents that represent the ball component. The ball component in the old implementation has been exaggerated for visualization purposes, while it is zero with the new implementation.

A test that captures the accuracy of the orientation conveyed by the vote is a comparison between the tangent of the ball vote and ground truth, which should be along the line connecting the voter and receiver. The old implementation was off by 1.25×10^{-5} degrees, while the new one was off by 1.38×10^{-5} degrees.

We also compared the two implementations in a simple experiment of local structure estimation. We sampled unoriented points from a sphere and a plane in 3D, and compared the estimated surface normals against the ground truth. Note that the coordinates of the points are quantized to make the evaluation fair for the old implementation, which only operates with integer coordinates, due to the data structure that stores the tokens. The inputs, which are encoded as unit balls, can be seen in Fig. 4.4(a). Surface and curve saliency maps for horizontal and vertical cuts of the data using both implementations are shown in Fig. 4.5. The saliency

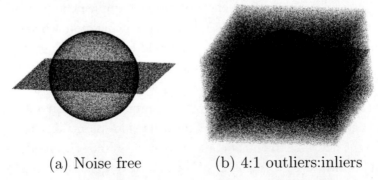

(a) Noise free   (b) 4:1 outliers:inliers

FIGURE 4.4: Sphere and plane inputs used for comparing the old and the new implementation of tensor voting.

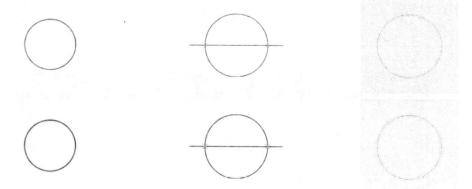

(a) Surface saliency z=120   (a) Surface saliency y=0   (a) Curve saliency z=0

FIGURE 4.5: Cuts of surface and curve saliency maps of the sphere and plane data. Darker areas correspond to higher saliency while white corresponds to zero. The top row was generated with the old implementation and the bottom row with the new one. The cuts of the curve saliency maps contain the plane and show that the intersection of the sphere and the plane is the most salient curve in the data.

maps are normalized so that their brightness covers the entire range from black to white. They are qualitatively almost identical for both cases. Quantitative results are presented in Table 4.1.

The accuracy in surface orientation estimation is similar in both cases, with a slight edge in favor of the new implementation. The main source of inaccuracy in the old implementation is the need for linear interpolation using the entries of the look-up table. It turns out that its effects on performance are similar to those of computing an approximate vote at the exact receiver position using the new approach. Since the difference in ball vote magnitudes becomes

TABLE 4.1: Results on the Sphere and Plane Dataset: Average Error Rate in Degrees for Normal Orientation Estimation Using the Implementation of [60] and the One Proposed Here

σ^2	OLD TV	NEW TV
50	2.24	1.60
100	1.47	1.18
200	1.09	0.98
500	0.87	0.93

TABLE 4.2: Results on the Sphere and Plane Dataset: Average Error Rate in Degrees for Normal Orientation Estimation Using the Implementation of [60] and the One Proposed Here

OUTLIERS:INLIERS	σ^2	OLD TV	NEW TV
1:1	50	3.03	2.18
	100	2.19	1.73
	200	1.75	1.48
	500	1.44	1.38
2:1	50	3.59	2.53
	100	2.61	2.02
	200	2.10	1.74
	500	1.74	1.62
5:1	50	4.92	3.36
	100	3.59	2.71
	200	2.90	2.33
	500	2.39	2.15
8:1	50	5.98	3.98
	100	4.33	3.20
	200	3.49	2.77
	500	2.89	2.62

important only when not all voters are ball tensors, we performed a second pass of tensor voting using the accumulated tensors from the first pass as voters. For $\sigma^2 = 100$, the error was 1.31° with the old implementation and 1.20° with the new one.

We, then, added noise to the data and repeated the experiment to test whether noise robustness is affected by the proposed approximation. The results are shown in the following table and are similar to the noise-free case. A safe conclusion is that noise robustness is not compromised by the new approximation.

It should also be noted here that the efficiency benefits of the new implementation become more apparent as the dimensionality increases. The inapplicability of the original implementation to high-dimensional datasets does not allow us to demonstrate the improvement quantitatively.

4.5   COMPUTER VISION PROBLEMS IN HIGH DIMENSIONS

In this section, we briefly review research on computer vision that has been addressed via tensor voting in high-dimensional spaces. By "high dimensional" here we refer to spaces with more than three dimensions. Results on machine learning problems with hundreds of dimensions are presented in the next chapter. The work presented here preceded the new efficient implementation, which would have achieved the same results faster.

4.5.1   Motion Analysis

Among the problems we addressed in high-dimensional spaces was perceptual organization in the absence of monocular information using visual motion cues only [73]. The representation is in the form of 4D tensors, since the goal is to enforce smoothness in the joint space of image coordinates and horizontal and vertical velocity. Salient solutions are characterized by smooth, in shape, image regions that move in a coherent way, thus receiving maximum support from all nearby pixels that have similar velocities.

After demonstrating the validity of our approach on synthetic data, we applied it to real images [72, 74, 75]. Candidate matches are generated by multiple cross correlation windows applied to all pixels, as in the case of stereo, but, since the epipolar constraint does not hold, the search for matches is done in 2D neighborhoods in the other image. The tokens are initialized as 4D ball tensors and tensor voting is performed to compute the saliency of the tokens. The token with the largest "surface" saliency ($\lambda_2 - \lambda_3$) is selected as the correct match for each pixel after outliers with low saliency are removed from the dataset. Since the input candidate matches are generated using correlation windows, foreground overextension described in Section 3.6 affects the quality of the results. Nicolescu and Medioni [75] proposed an edge-based method to correct overextensions. The initial motion boundaries estimated by tensor voting are used to define areas where the actual boundaries are, under the assumption that the errors are at most equal to the width of the occluded regions plus half the width of the matching windows. An edge detector is applied to these areas, taking into account the estimated orientation of the boundaries, and the most salient edges are detected and used to trim the overextensions in the velocity maps. Results can be seen in Fig. 4.6.

4.5.2   Epipolar Geometry Estimation

A parameter estimation problem that occurs often in computer vision is epipolar geometry estimation from putative pixel correspondences in two images. The epipolar geometry is described by the fundamental matrix, which is a 3×3 matrix in augmented space and thus has eight degrees of freedom. Therefore, one needs eight correct correspondences to obtain the fundamental matrix using linear estimation. Typically, iterative random sampling algorithms such as RANSAC are applied followed by nonlinear refinement. See [27] for an excellent treatment.

(a) Input image (b) Initial matches

(c) Initial boundaries (d) Corrected boundaries

FIGURE 4.6: Results on the "candybox" example. The initial matches are shown rotated in (x, y, v_x) space. The second row shows results before discontinuity localization and after the edge-based correction.

A different, noniterative approach was proposed by Tang, Medioni, and Lee [105, 106]. The problem of finding a structure with eight degrees of freedom is posed as finding the most dominant hyperplane in an 8D space in which the constraint for the fundamental matrix provided by each putative pixel correspondence is represented as a point. Correct matches align in this space to form a hyperplane that becomes salient among the clutter generated by erroneous matches. Similar parameter estimation problems, such as the estimation of projection matrices, 2D and 3D homographies, and other transformations that are defined by sets of inliers corrupted by numerous outliers, can be addressed in the same way. Each inlier provides one or more linear constraints on the parameters of the transformation. The points can be viewed as points in a high-dimensional space, while the transformations as salient structures comprised of these points. Tensor voting provides a noniterative, model-free method for finding these structures, as well as estimating the actual number of degrees of freedom. The latter can help in the detection of degenerate cases.

The problem of estimating the epipolar geometry given tentative pixel correspondences in two images was later attacked from a different angle by Tong, Tang, and Medioni in [110, 111].

This approach exploits the observation that correct correspondences form cones in the 4D joint image space, where the coordinates are the concatenation of the coordinates of the two images. Independently moving objects generate different cones and follow separate epipolar geometries that are defined by the relative motion of the camera and each object. All available epipolar geometries can be recovered by extracting all salient cones in the joint image space.

4.5.3   Texture Synthesis

A computer vision application in high dimensions was presented by Jia and Tang in [35, 36], where they address texture synthesis using tensor voting. Processing begins by identifying regions

(a) Input image

(b) Removed sign

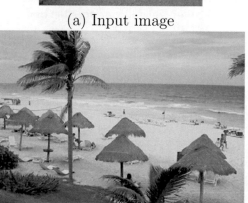

(c) Input image

(d) Removed palm tree

FIGURE 4.7: Results on the "signpost" and "beach" images from [35] where the sign and a palm tree, respectively, have been removed and replaced by synthesized texture that matches the background.

in the image where the texture has to be seamlessly replaced with texture from neighboring regions. The boundaries of these regions are extrapolated in the region where the synthesis will take place in a way that looks perceptually plausible. Then, the missing texture is synthesized via tensor voting in an ND space, where N is the size of the window that is used to represent "textons". It is automatically chosen to match the properties of the input image. For a value of N equal to 7, for instance, textons are represented by 7×7 windows and new textons are synthesized by tensor voting in a 50D space, where the first 49 dimensions are the elements of the square window stacked in a vector and the fiftieth dimension is the maximum intensity. Representative results on object removal from challenging images are shown in Fig. 4.7.

4.6   DISCUSSION

In this chapter, we have presented a significant contribution to the tensor voting framework. It allows us to apply our methodology to problems where the properties of tensor voting, such as its noise robustness and lack of global computations, seemed appropriate and desirable. However, the faithful adherence to exact vote generation made computational and storage requirements impractical. Experiments on data with ground truth show that the new approximation is equally effective in orientation estimation and also maintains the noise robustness of the original implementation. These results suggest that the main useful source of uncertainty in the tensors comes as a result of the tensor addition of votes at each token, and not from the uncertainty component of each vote, the computation of which we bypass here.

We have also, very briefly, presented some earlier work on computer vision problems with dimensionality higher than 3. Even though this research was carried out with previous high-dimensional implementations of tensor voting, it shows that there are numerous computer vision applications that can be addressed as perceptual organization in spaces beyond the 2D image domain or the 3D Euclidean world.

In the following chapter, we demonstrate outstanding results in problems such as dimensionality estimation, manifold learning, nonlinear interpolation, geodesic distance measurement, and function approximation. We anticipate that the work presented here will serve as the groundwork for research in domains that include instance-based learning, pattern recognition, classification, data mining, and kinematics.

CHAPTER 5

Dimensionality Estimation, Manifold Learning and Function Approximation

Machine learning is a research area in artificial intelligence that aims at the improvement of the behavior of agents through diligent study of observations [84]. It deals with the development of algorithms that analyze the observed data to identify patterns and relationships, in order to predict unseen data. Here, we address a subfield of machine learning that operates in continuous domains and learns from observations that are represented as points in a Euclidean space. This type of learning is termed *instance-based* or *memory-based* learning [62]. Learning in discrete domains, which attempts to infer the states, transitions and rules that govern a system, or the decisions and strategies that maximize a utility function, is out of the scope of our research.

The problem of learning a target function based on instances is equivalent to learning a manifold formed by a set of points, and thus being able to predict the positions of other points on the manifold. The first task, given a set of observations, is to determine the intrinsic dimensionality of the data. This can provide insight for the complexity of the system that generates the data, the type of model needed to describe them, as well as the actual degrees of freedom of the system, which are not necessarily equal with the dimensionality of the input space. We also estimate the orientation of a potential manifold that passes through each point by tensor voting.

Instance-based learning has recently received renewed interest from the machine learning community, due to its many applications in the fields of pattern recognition, data mining, kinematics, function approximation and visualization, among others. This interest was sparked by a wave of new algorithms that advanced the state of the art and are capable of learning nonlinear manifolds in spaces of very high dimensionality. These include kernel PCA [98], locally linear embedding (LLE) [82], Isomap [109] and charting [10], which are reviewed in Section 5.1. They aim at reducing the dimensionality of the input space in a way that preserves certain geometric or statistical properties of the data. Isomap, for instance, attempts to preserve the geodesic distances between all points as the manifold is "unfolded" and mapped to a space of lower dimension. A common assumption is that the desired manifold consists of locally linear patches. We relax this assumption by only requiring that manifolds be smooth almost everywhere.

We take a different path to learning low dimensional manifolds from instances in a high dimensional space. Whereas traditional methods address the problem as one of dimensionality reduction, we propose an approach for the unsupervised learning of manifold structure in a way that is useful for tasks such as geodesic distance estimation and nonlinear interpolation, that does not embed the data in a lower dimensional space. We compute local dimensionality estimates, but instead of performing dimensionality reduction, we perform all operations in the original input space, taking into account the estimated dimensionality of the data. This allows us to process datasets that are not manifolds globally, or ones with varying intrinsic dimensionality. The latter pose no additional difficulties, since we do not use a global estimate for the dimensionality of the data. Results have been presented in [69]. Moreover, outliers, boundaries, intersections or disconnected components are handled naturally as in 2-D and 3-D. Non-manifolds, such as hyper-spheres, can also be processed without any modifications of the algorithm since we do not attempt to estimate a global "unfolding". Quantitative results for the robustness against outliers that outnumber the inliers are presented in Sections 5.3 and 5.4.

Manifold learning serves as the basis for the last part of our research, which addresses function approximation. The approximation of an unknown function based on observations is critical for predicting the responses of both natural and artificial systems. The main assumption is that some form of smoothness exists in the data [80] and unobserved outputs can be predicted from previously observed outputs for similar inputs. The distinction between low and high-dimensional spaces is necessary, since highly specialized methods for low-dimensional cases exist in the literature. Here, we address the approximation of multivariate functions, and thus employ methods that can be generalized to high dimensions. A common practice is to treat functions with multiple outputs as multiple single-output functions. We adopt this scheme here, even though nothing prohibits us from directly approximating multiple-input multiple-output functions.

As suggested by Poggio and Girosi [80], function approximation from samples and hypersurface inference are equivalent. Our approach is local, nonparametric and has a weak prior model of smoothness. The fact that no model other than local constant curvature connections between the voter and receiver is used allows us to handle a broad range of functions. Also contributing to this is the absence of global computations and parameters, such as the number of local models in an ensemble, that need to be selected. The inevitable trade-off between over-smoothing and over-fitting is regulated by the selection of the scale, which is equivalent to the radius of the voting neighborhood. Small values reduce the size of the voting neighborhood and preserve details better, but are more vulnerable to noise and over-fitting. Large values produce smoother approximations that are more robust to noise. As shown in Section 5.5, the results are very stable with respect to the scale. As most of the local methods reviewed in the next section, our algorithm is memory based. This increases flexibility, since we can process data that do not conform to any model, but also increases storage requirements, since all samples are kept in memory.

This chapter is organized as follows: an overview of related work including the algorithms that are compared with ours is given in the next section; results in dimensionality estimation are presented in Section 5.2, while results in local structure estimation are presented in Section 5.3; our algorithm for measuring distances on the manifold and a quantitative comparison with state of the art methods is presented in Section 5.4; an algorithm for generating outputs for unobserved inputs that is used for function approximation is described in Section 5.5; finally, Section 5.6 concludes the chapter.

5.1   RELATED WORK

In this section, we present related work in the domains of dimensionality estimation, manifold learning and multivariate function approximation.

Dimensionality Estimation.   Bruske and Sommer [13] present an approach for dimensionality estimation where an optimally topology preserving map (OTPM) is constructed for a subset of the data after vector quantization. Principal Component Analysis (PCA) [37] is then performed for each node of the OTPM under the assumption that the underlying structure of the data is locally linear. The average of the number of significant singular values at the nodes is the estimate of the intrinsic dimensionality.

Kégl [41] estimates the capacity dimension of the manifold, which does not depend on the distribution of the data, and is equal to the topological dimension, using an efficient approximation based on packing numbers. Costa and Hero [14] estimate the intrinsic dimension of the manifold and the entropy of the samples using geodesic-minimal-spanning trees. The method, similarly to Isomap [109], considers global properties of the adjacency graph and thus produces a single global estimate. Levina and Bickel [52] compute maximum likelihood estimates of dimensionality by examining the number of neighbors included in spheres the radius of which is selected in such a way that they contain enough points and that the density of the data contained in them can be assumed constant. These requirements cause an underestimation of the dimensionality when it it very high.

Manifold Learning.   Here, we briefly present recent approaches for learning low dimensional embeddings from points in high dimensional spaces. Most of them are extensions of linear techniques, such as Principal Component Analysis (PCA) [37] and Multi-Dimensional Scaling (MDS) [15], based on the assumption that nonlinear manifolds can be approximated by locally linear patches.

In contrast to other methods, Schölkopf et al. [98] propose kernel PCA that attempts to find linear patches using PCA in a space of typically higher dimensionality than the input space. Correct kernel selection can reveal the low dimensional structure of the input data after mapping the instances to a space of higher dimensionality. For instance a second order polynomial kernel can detect quadratic surfaces since they appear as planes in the high-dimensional space.

Locally Linear Embedding (LLE) was presented by Roweis and Saul [82, 90]. The underlying assumption is that if data lie on a locally linear, low-dimensional manifold, then each point can be reconstructed from its neighbors with appropriate weights. These weights should be the same in a low-dimensional space, the dimensionality of which is greater or equal to the intrinsic dimensionality of the manifold, as long as the manifold is locally linear. The LLE algorithm computes the basis of such a low-dimensional space. The dimensionality of the embedding, however, has to be given as a parameter, since it cannot always be estimated from the data [90]. Moreover, the output is an embedding of the given data, but not a mapping from the ambient to the embedding space. Global coordination of the local embeddings, and thus a mapping, can be computed according to [108]. LLE is not isometric and often fails by mapping distant points close to each other.

Tenenbaum et al. [109] propose Isomap, which is an extension of MDS that uses geodesic instead of Euclidean distances. This allows Isomap to handle nonlinear manifolds, whereas MDS is limited to linear data. The geodesic distances between points are approximated by graph distances. Then, MDS is applied on the geodesic distances to compute an embedding that preserves the property of points to be close or far away from each other. Due to its global formulation, Isomap's computational cost is considerably higher than that of LLE. The benefit is that not only it preserves distances between nearest neighbors, but between all points. In addition, it can handle points not in the original dataset, and perform interpolation. C-Isomap, a variation of Isomap that can be applied to data with intrinsic curvature, but known distribution, and L-Isomap, a faster alternative that only uses a few landmark point for distance computations, have also been proposed in [16]. Isomap and its variants are limited to convex datasets.

The Laplacian Eigenmaps algorithm was developed by Belkin and Niyogi [6]. It computes the normalized graph Laplacian of the adjacency graph of the input data, which is an approximation of the Laplace-Beltrami operator on the manifold. It exploits locality preserving properties that were first observed in the field of clustering. The Laplacian Eigenmaps algorithm can be viewed as a generalization of LLE, since the two become identical when the weights of the graph are chosen according to the criteria of the latter. Much like LLE, the dimensionality of the manifold also has to be provided, the computed embeddings are not isometric and a mapping between the two spaces is not produced. The latter is addressed in [28] where a variation of the algorithm is proposed.

Donoho and Grimes [18] propose Hessian LLE (HLLE), an approach similar to the above, which computes the Hessian instead of the Laplacian of the graph. The authors claim that the Hessian is better suited than the Laplacian for detecting linear patches on the manifold. The major contribution of this approach is that it proposes a global, isometric method, which, unlike Isomap, can be applied to non-convex datasets. The need to estimate second derivatives from possibly noisy, discrete data makes the algorithm more sensitive to noise than the others reviewed here.

Semidefinite Embedding (SDE) was proposed by Weinberger and Saul [117] who address the problem of manifold learning by enforcing local isometry. The lengths of the sides of triangles formed by neighboring points are preserved during the embedding. These constraints can be expressed in terms of pairwise distances and the optimal embedding can be found by semidefinite programming. The method is the most computationally demanding reviewed here. However, it can reliably estimate the underlying dimensionality of the inputs by locating the largest gap between the eigenvalues of the Gram matrix of the outputs. Similarly to our approach, this estimate does not require a threshold.

Other research related to ours includes the charting algorithm of Brand [10]. It computes a pseudo-invertible mapping of the data, as well as the intrinsic dimensionality of the manifold. The latter is estimated by examining the rate of growth of the number of points contained in hyper-spheres as a function of the radius. Linear patches, areas of curvature and noise can be discriminated against using the proposed measure. Affine transformations that align the coordinate systems of the linear patches are computed at the second stage. This defines a global coordinate system for the embedding and thus a mapping between the input space and the embedding space.

Wang *et al.* [116] propose an adaptive version of the local tangent space alignment (LTSA) of Zhang and Zha [127], a local dimensionality reduction method that is a variation of LLE. Wang *et al.* address a limitation of most of the approaches presented in this section, which is the use of a fixed number of neighbors for all points in the data. This causes serious problems if that number is not selected properly, for points near boundaries, or if the density of the data is not approximately constant.

The difference between our approach and those of [10, 13, 14, 41, 52, 117] is that ours produces reliable dimensionality estimates at the point level, which do not have to be averaged over the entire dataset. While this is not important for datasets with constant dimensionality, the ability to estimate local dimensionality reliably becomes a key factor when dealing with data generated by different unknown processes. Given reliable local estimates, the dataset can be segmented in components with constant dimensionality.

Function Approximation.   Neural networks are often employed as global methods for function approximation. Poggio and Girosi [80] addressed function approximation in a regularization framework implemented as a three-layer neural network. They view the problem as hypersurface reconstruction, where the only reasonable assumption is that of smoothness. The emphasis is on the selection of the appropriate approximating functions and optimization algorithm. Other global methods include the work of Sanger [87], Barron [3], Breiman [11], Saha *et al.*[85], Xu *et al.* [122] and Mitaim and Kosko [61].

Lawrence *et al.* [48] compared a global approach using a multi-layer perceptron neural network with a linear local approximation model. They found that the local model performed better when the density of the input data deviated a lot from being uniform. Furthermore, the local model allowed for incremental learning and cross-validation. On the other hand, it showed poorer generalization, slower performance after training and required more memory, since all input data had to be stored. The global model performed better in higher dimensions, where data sparsity becomes a serious problem for the local alternative.

Schaal and Atkenson [94] proposed a nonparametric, local, incremental learning approach based on receptive field weighted regression. The approach is truly local since the parameters for each model and the size and shape of each receptive field are learned independently. The provided mechanisms for the addition and pruning of local models enable incremental learning as new data points become available. Atkenson *et al.* [2] survey local weighted learning methods and identify the issues that must be taken into account. These include the selection of the distance metric, the weighting function, prediction assessment and robustness to noise. The authors argue that in certain cases no values of the parameters of a global model can provide a good approximation of the true function. In these cases, a local approximation using a simpler, even linear model, is a better approach than increasing the complexity of the global model. Along these lines, Vijaykumar and Schaal [115] proposed locally weighted projection regression, an algorithm based on successive univariate regressions along projections of the data in directions given by the gradient of the underlying function.

We also opt for a local approach and address the problem as manifold learning. Note, however, that we are not limited to functions that are strictly manifolds. The recent group of manifold learning algorithms based on dimensionality reduction, with the exception of the adaptive local tangent space alignment method [116], is not applicable for function approximation. This is because they compute neighborhood relationships in the form of a graph, but do not compute the geometric structure of the observations and thus cannot generate new, unobserved instances on the manifold. Using tensor voting, we are able to reliably estimate the normal and tangent space at each sample, as described in the following section. These estimates allow us to perform nonlinear interpolation and generate outputs for unobserved inputs, even under severe noise corruption. Since the votes are weighted, sensitivity to the scale of voting and outliers is small, as demonstrated by the experiments in the remainder of the paper.

FIGURE 5.1: The "Swiss Roll" dataset in 3-D

5.2   DIMENSIONALITY ESTIMATION

In this section, we present experimental results in dimensionality estimation. As described in Section 4.3.3, the intrinsic dimensionality at each point can be found as the maximum gap in the eigenvalues of the tensor after votes from its neighboring points have been collected. All inputs consist of unoriented points and are encoded as ball tensors.

Swiss Roll.   The first experiment is on the "Swiss Roll" dataset, which is available online at http://isomap.stanford.edu/. It contains $20,000$ points on a 2-D manifold in 3-D (Fig. 5.1). We perform a simple evaluation of the quality of the orientation estimates by projecting the nearest neighbors of each point on the estimated tangent space and measuring the percentage of the distance that has been recovered. This is a simple measure of the accuracy of the local linear approximation of the nonlinear manifold. The percentage of points with correct dimensionality estimates and the percentage of recovered distances for the 8 nearest neighbors as a function of σ, can be seen in Table 5.1. The performance is the same at boundaries, which do not pose any additional difficulties to our algorithm. The number of votes cast by each point ranges from 187

TABLE 5.1: Rate of correct dimensionality estimation and execution times as functions of σ for the "Swiss Roll" dataset.

σ	CORRECT DIM. ESTIMATION (%)	PERC. OF DIST. RECOVERED (%)	TIME (SEC)
50	99.25	93.07	7
100	99.91	93.21	13
200	99.95	93.19	30
300	99.92	93.16	47
500	99.68	93.03	79
700	99.23	92.82	112
1000	97.90	92.29	181

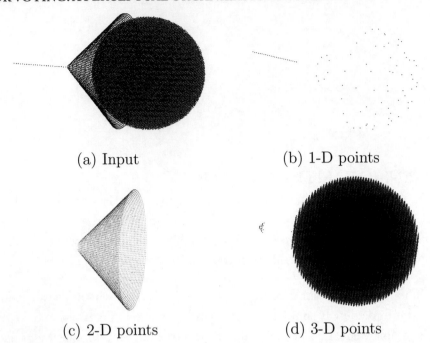

(a) Input   (b) 1-D points

(c) 2-D points   (d) 3-D points

FIGURE 5.2: Data of varying dimensionality in 4-D. The first three axes of the input and the classified points are shown. Note that the hyper-sphere is empty in 4-D, but appears as a full sphere when visualized in 3-D.

for $\sigma = 50$ to 5440 for $\sigma = 1000$. The reported processing times are for a Pentium 4 processor at 2.8 GHz. A conclusion that can safely be drawn from the table is that the accuracy is high and stable for a large range of values of σ.

Structures with Varying Dimensionality.   The second dataset is in 4-D and contains points sampled from three structures: a line, a 2-D cone and a 3-D hyper-sphere. The hyper-sphere is a structure with three degrees of freedom. It cannot be unfolded unless we remove a small part from it. Figure 5.2(a) shows the first three dimensions of the data. The dataset contains a total 135, 864 points, which are encoded as ball tensors. Tensor voting is performed with $\sigma = 200$. Figures 5.2(b-d) show the points classified according to their dimensionality. Performing the same analysis as above for the accuracy of the tangent space estimation, 91.04% of the distances of the 8 nearest neighbors of each point lie on the tangent space, even though both the cone and the hyper-sphere have intrinsic curvature and cannot be accurately approximated by linear models. All the methods presented in Sec. 5.1 fail for this dataset because of the presence of structures with different dimensionalities and because the hyper-sphere cannot be unfolded.

TABLE 5.2: Rate of correct dimensionality estimation for high dimensional data

INTRINSIC DIM.	LINEAR MAPPINGS	QUADRATIC MAPPINGS	SPACE DIM.	DIM. EST. (%)
4	10	6	50	93.6
3	8	6	100	97.4
4	10	6	100	93.9
3	8	6	150	97.3

Data in High Dimensions.   The datasets for this experiment were generated by sampling a few thousand points from a low-dimensional space (3- or 4-D) and mapping them to a medium dimensional space (14- to 16-D) using linear and quadratic functions. The generated points were then rotated and embedded in a 50- to 150-D space, while outliers drawn from a uniform distribution were added to the dataset. The percentage of correct point-wise dimensionality estimates after tensor voting can be seen in Table 5.2.

5.3   MANIFOLD LEARNING

In this section, we present quantitative results on simple datasets in 3-D for which ground truth can be analytically computed. In Section 5.4, we process the same data with state of the art manifold learning algorithms and compare their results against ours. The two datasets are a section of a cylinder and a section of a sphere shown in Fig. 5.3. The cylindrical section spans 150° and consists of 1000 points. The spherical section spans 90° × 90° and consists of 900 points. Both are approximately uniformly sampled. The points are represented by ball tensors, assuming no information about their orientation. In the first part of the experiment, we compute

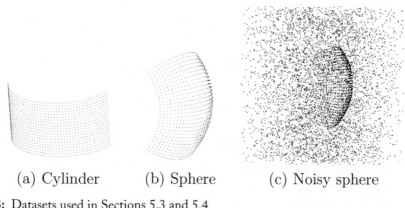

(a) Cylinder   (b) Sphere   (c) Noisy sphere

FIGURE 5.3: Datasets used in Sections 5.3 and 5.4

TABLE 5.3: Results on the cylinder dataset. Shown in the first column is σ, in the second is the average number of neighbors that cast votes to each point, in the third the average error in degrees of the estimated normals, and in the fourth the accuracy of dimensionality estimation.

σ	NEIGHBORS	ANGULAR ERROR	DIM. ESTIM. (%)
10	5	0.06	4
20	9	0.07	90
30	9	0.08	90
40	12	0.09	90
50	20	0.10	100
60	20	0.11	100
70	23	0.12	100
80	25	0.12	100
90	30	0.13	100
100	34	0.14	100

local dimensionality and normal orientation as a function of scale. The results are presented in Tables 5.3 and 5.4. The results show that if the scale is not too small, dimensionality estimation is very reliable. For all scales the angular errors are below $0.4°$. Similar results are obtained for a large range of scales.

The same experiments were performed for the spherical section in the presence of outliers. Quantitative results are shown in the following tables for a number of outliers that ranges from 900 (equal to the inliers) to 5000. The latter dataset is shown in Fig. 5.3(c). Note that performance was evaluated only on the points that belong to the sphere and the results are shown in Table 5.5. Larger values of the scale prove to be more robust to noise, as expected. The smallest values of the scale result in voting neighborhoods that include less than 10 points, which are insufficient. Taking this into account, performance is still good even with wrong parameter selection. Also note that one could reject the outliers by thresholding, since they have smaller eigenvalues than the inliers, and perform tensor voting again to obtain even better estimates of structure and dimensionality. Even a single pass of tensor voting, however, turns out to be very effective, especially considering that no other method can handle such a large number of outliers. Foregoing the low-dimensional embedding is a main reason that allows our method to perform well in the presence of noise, since embedding random outliers in a low-dimensional space would make their influence more detrimental. This is due to the structure imposed to them by the mapping, which makes the outliers less random. It is also due to the increase in

TABLE 5.4: Results on the sphere dataset. The columns are the same as in Table 5.3.

σ	NEIGHBORS	ANGULAR ERROR	DIM. ESTIM. (%)
10	5	0.20	44
20	9	0.23	65
30	11	0.24	93
40	20	0.26	94
50	21	0.27	94
60	23	0.29	94
70	26	0.31	94
80	32	0.34	94
90	36	0.36	94
100	39	0.38	97

their density in the low-dimensional space compared to that in the original high-dimensional space.

5.4 MANIFOLD DISTANCES AND NONLINEAR INTERPOLATION

Learning the manifold structure from samples is an interesting problem. The ability to evaluate intrinsic manifold distances between points and to interpolate on the manifold are more useful for many applications. Here, we show how to compute the distance between any two points on a manifold essentially by taking small steps on the manifold, collecting votes, estimating the local tangent space and advancing on it until the destination is reached.

Processing begins by learning the manifold structure, as in the previous section, starting from unoriented points that are represented by ball tensors. After tensor voting we obtain dimensionality and orientation estimates. We can travel on the manifold by selecting a starting point, which has to be on the manifold, and a target point or a desired direction (the vector from the origin to the target). At each step, we can project the desired direction on the tangent space of the current point and create a new point a small distance away. The tangent space of the new point is computed by collecting votes from the neighboring points, as in regular tensor voting. Note that the tensors used here are no longer balls, but the ones resulting from the previous pass of voting. The process is illustrated in Fig. 5.4, where we start from point A and wish to reach B. We project $\vec{t}$, the vector from A to B, on the estimated tangent space of A and obtain its projection $\vec{p}$. Then, we take a small step along $\vec{p}$ to point A_1, on which we

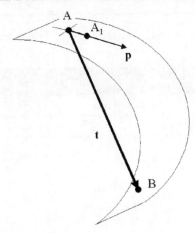

FIGURE 5.4: Nonlinear interpolation on the tangent space of a manifold

collect votes to obtain an estimate of its tangent space. The desired direction is then projected on the tangent space of the new point and so forth until the destination is reached within ϵ. The manifold distance between A and B is approximated by measuring the length of the path. In the process, we have also generated a number of new points on the manifold, which may be a desirable by-product for some applications.

TABLE 5.5: Results on the sphere dataset contaminated by noise. AE: error in normal angle estimation in degrees, DE: correct dimensionality estimation (%).

OUTLIERS	900		3000		5000	
σ	AE	DE	AE	DE	AE	DE
10	1.15	44	3.68	41	6.04	39
20	0.93	65	2.95	52	4.73	59
30	0.88	92	2.63	88	4.15	85
40	0.88	93	2.49	90	3.85	88
50	0.90	93	2.41	92	3.63	91
60	0.93	94	2.38	93	3.50	93
70	0.97	94	2.38	93	3.43	93
80	1.00	94	2.38	94	3.38	94
90	1.04	95	2.38	95	3.34	94
100	1.07	97	2.39	95	3.31	95

TABLE 5.6: Error rates in distance measurements between pairs of points on the manifolds. The best result of each method is reported along with the number of neighbors used for the embedding (K), or the scale σ in the case of tensor voting (TV).

DATASET	SPHERE		CYLINDER	
	K	ERR(%)	K	ERR(%)
LLE	18	5.08	6	26.52
Isomap	6	1.98	30	0.35
Laplacian	16	11.03	10	29.36
HLLE	12	3.89	40	26.81
SDE	2	5.14	6	25.57
TV (σ)	60	0.34	50	0.62

The first experiment on manifold distance estimation is a quantitative evaluation against some of the most widely used algorithms of the literature. For the results reported in Table 5.6, we learn the local structure of the manifolds of the previous section using tensor voting. We also compute embeddings using LLE [82], Isomap [109], Laplacian eigenmaps [6], HLLE [18] and SDE [117]. Matlab implementations for these methods can be downloaded from the following internet locations.

- LLE from http://www.cs.toronto.edu/~roweis/lle/code.html
- Isomap from http://isomap.stanford.edu/
- Laplacian Eigenmaps from http://people.cs.uchicago.edu/~misha/ManifoldLearning/index.html
- HLLE from http://basis.stanford.edu/HLLE and
- SDE from http://www.seas.upenn.edu/~kilianw/sde/download.htm.

We are grateful to the authors for making the core of their methods available to the community. We intend to make our software publicly available as well.

The experiment is performed as follows. We randomly select 5000 pairs of points on each manifold and attempt to measure the geodesic distance between the points of each pair in the input space using tensor voting and in the embedding space using the other five methods. The estimated distances are compared to the ground truth: $r\Delta\theta$ for the sphere and $\sqrt{(r\Delta\theta)^2 + (\Delta z)^2}$ for the cylinder. Among the above approaches, only Isomap and SDE produce isometric embeddings, and only Isomap preserves the absolute distances between the input and the embedding space. To make the evaluation fair, we compute a uniform scale that minimizes the error between the computed distances and the ground truth for all methods, except Isomap for which it is not necessary. Thus, perfect distance ratios would be awarded a perfect rating in the evaluation, even if the absolute magnitudes of the distances are meaningless in the embedding space. For all the algorithms, we tried a wide range for the number of neighbors, K. In some cases, we were not able to produce good embeddings of the data for any value of K, especially for the cylinder. Given the fact that we scale the data, errors above 20% indicate very poor performance, which is also confirmed by visual inspection of the embeddings.

The evaluation of the quality of manifold learning based on the computation of pairwise distances is a fair measure for the performance of all algorithms, since high quality manifold learning should minimize distortions. In addition, the proposed evaluation does not require operations that are not supported by some of the algorithms, such as the processing of points not included in the training dataset. Quantitative results are presented in Table 5.6 along with the value of K that is used. In the case of tensor voting, the same scale is used for both learning the manifold and computing distances.

We also apply our method in the presence of 900, 3000 and 5000 outliers. Keep in mind that the sphere and the cylinder datasets consist of 900 and 1000 points respectively. The error rates using tensor voting for the sphere are 0.39%, 0.47% and 0.53% respectively. The rates for the cylinder are 0.77%, 1.17% and 1.22%. Compared with the noise free case, these results demonstrate that our approach degrades slowly in the presence of outliers. The best performance achieved by any other method is 3.54% on the sphere dataset with 900 outliers by Isomap. Complete results are shown in Table 5.7. In many cases, we were unable to achieve useful embeddings for datasets with outliers. The results using tensor voting can be found in Table 5.8.

Datasets with Varying Dimensionality and Intersections.   For the final experiment of this section, we create synthetic data in 3-D that were embedded in higher dimensions. The first dataset consists of a line and a cone. The points are embedded in 50-D by three orthonormal 50-D vectors and initialized as ball tensors. Tensor voting is performed in the 50-D space and a path from point A on the line to point B on the cone is interpolated as in the previous experiment, making sure that it belongs to the local tangent space, which changes dimensionality from one to two. The data is re-projected back to 3-D for visualization in Fig. 5.5(a).

In the second part of the experiment, we generate an intersecting S-shaped surface and a plane (a total of 11,000 points) and 30,000 outliers, and embed them in a 30-D space. Without explicitly removing the noise, we interpolate between two points on the S (A and B) and a point on the S and a point on the plane (C and D) and create the paths shown in Fig. 5.5(b)

TABLE 5.7: Error rates in distance measurements between pairs of points on the manifolds under outlier corruption. The best result of each method is reported along with the number of neighbors used for the embedding (K), or the scale σ in the case of tensor voting (TV). Note that HLLE fails to compute an embedding for small values of K, while SDE fails at both examples for all choices of K.

	SPHERE		CYLINDER	
DATASET	900 K	OUTLIERS ERR(%)	900 K	OUTLIERS ERR(%)
LLE	40	60.74	6	15.40
Isomap	18	3.54	14	11.41
Laplacian	6	13.97	14	27.98
HLLE	30	8.73	30	23.67
SDE		N/A		N/A
TV (σ)	70	0.39	100	0.77

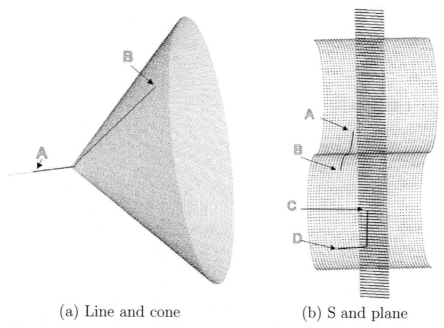

(a) Line and cone   (b) S and plane

FIGURE 5.5: Nonlinear interpolation in 50-D with varying dimensionality (a) and 30-D with intersecting manifolds under noise corruption (b).

TABLE 5.8: Error rates for our approach in the presence of 3000 and 5000 outliers

DATASET	σ	ERROR RATE
Sphere (3000 outliers)	80	0.47
Sphere (5000 outliers)	100	0.53
Cylinder (3000 outliers)	100	1.17
Cylinder (5000 outliers)	100	1.22

re-projected in 3-D. The first path is curved, while the second jumps from manifold to manifold still keeping the optimal path. (The outliers are not shown for clarity.) Processing time for 41,000 points in 30-D is 2 min. and 40 sec. on a Pentium 4 at 2.8 MHz using voting neighborhoods that included an average of 44 points.

5.5   GENERATION OF UNOBSERVED SAMPLES AND NONPARAMETRIC FUNCTION APPROXIMATION

In this section, we build upon the results of the previous section to address function approximation. As before, observations of inputs and outputs are available for training. The difference with the examples of the previous sections is that the queries are given as input vectors with unknown output values, and thus are of lower dimension than the voting space. The missing module is one that can find a point on the manifold that corresponds to an input similar to the query. Then, in order to predict the output y of the function for an unknown input $\vec{x}$, under the assumption of local smoothness, we move on the manifold formed by the training samples until we reach the point corresponding to the given input coordinates. To ensure that we always remain on the manifold, we need to start from a point on it and proceed as in the previous section.

One way to find a suitable starting point is to find the nearest neighbor of $\vec{x}$ in the input space, which has fewer dimensions than the joint input-output (voting) space. Then, we can compute the desired direction in the low dimensional space and project it to the input-output space. If many outputs are possible for a given input (if the data have not been generated by a function in the strict sense), we have to either find neighbors at each branch of the function and produce multiple outputs, or use other information, such as the previous state of the system, to pursue only one of the alternatives. Figure 5.6 provides a simple illustration. We begin with a point A_i in the input space. We proceed by finding its nearest neighbor among the projections of the training data on the input space B_i. (Even if B_i is not the nearest neighbor the scheme still works but possibly requires more steps.) The sample B in the input-output space that corresponds to B_i is the starting point on the manifold. The desired direction is the projection

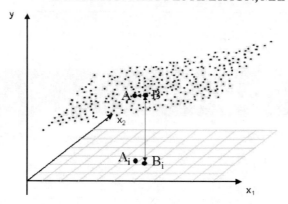

FIGURE 5.6: Interpolation to obtain output value for unknown input point A_i

of the $A_i B_i$ vector on the tangent space of B. Now, we arc in the case described in Section 5.4, where the starting point and the desired direction are known. Processing stops when the input coordinates of the point on the path from B are within ϵ of A_i. The corresponding point A in the input-output space is the desired interpolated sample.

As in all the experiments presented in this paper, the input points are encoded as ball tensors, since we assume that we have no knowledge of their orientation. The first two experiments we conducted were on functions proposed in [116]. The key difficulty with these functions is the non-uniform density of the data. In the first example we attempt to approximate:

$$x_i = [\cos(t_i), \sin(t_i)]^T \qquad t_i \in [0, \pi], \, t_{i+1} - t_i = 0.1(0.001 + |\cos(t_i)|) \qquad (5.1)$$

where the distance between consecutive samples is far from uniform. See Fig. 5.7(a) for the inputs and the second column of Table 5.9 for quantitative results on tangent estimation for 152 points as a function of scale.

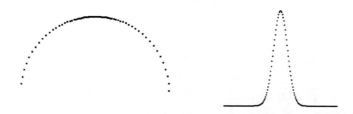

(a) Samples from Eq. 5.1   (b) Samples from Eq. 5.2

FIGURE 5.7: Input data for the two experiments proposed by [116]

TABLE 5.9: Error in degrees for tangent estimation for the functions of Eq. 5.1 and Eq. 5.2

σ	EQ. 5.1 152 POINTS	EQ. 5.2 180 POINTS	EQ. 5.2 360 POINTS
10	0.60	4.52	2.45
20	0.32	3.37	1.89
30	0.36	2.92	1.61
40	0.40	2.68	1.43
50	0.44	2.48	1.22
60	0.48	2.48	1.08
70	0.51	2.18	0.95
80	0.54	2.18	0.83
90	0.58	2.02	0.68
100	0.61	2.03	0.57

In the second example, which is also taken from [116], points are uniformly sampled on the t-axis from the $[-6, 6]$ interval. The output is produced by the following function:

$$x_i = [t_i, \quad 10e^{-t_i^2}] \tag{5.2}$$

The points, as can be seen in Fig. 5.7(b), are not uniformly spaced. The quantitative results on tangent estimation accuracy for 180 and 360 samples from the same interval are reported in the last two columns of Table 5.9. Naturally, as the sampling becomes denser, the quality of the approximation improves. What should be emphasized here is the stability of the results as a function of σ. Even with as few as 5 or 6 neighbors included in the voting neighborhood, the tangent at each point is estimated quite accurately.

For the next experiment we approximate the following function, proposed by Schaal and Atkenson [94]:

$$y = max\{e^{-10x_1^2}, \quad e^{-50x_2^2}, \quad 1.25e^{-5(x_1^2+x_2^2)}\} \tag{5.3}$$

1681 samples of y are generated by uniformly sampling the $[-1, 1] \times [-1, 1]$ square. We perform four experiments with increasing degree of difficulty. In all cases, after voting on the given inputs, we generate new samples by interpolating between the input points. The four configurations and noise conditions were:

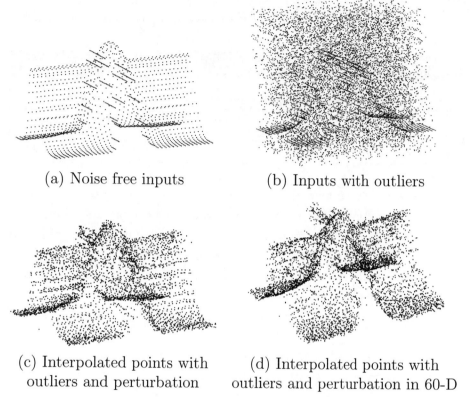

(a) Noise free inputs   (b) Inputs with outliers

(c) Interpolated points with   (d) Interpolated points with
outliers and perturbation   outliers and perturbation in 60-D

FIGURE 5.8: Inputs and interpolated points for Eq. 5.3. The top row shows the noise-free inputs and the noisy input set where only 20% of the points are inliers. The bottom row shows the points generated in 3-D and 60-D respectively. In both cases the inputs were contaminated with outliers and Gaussian noise.

- In the first experiment, we performed all operations with noise free data in 3-D.
- For the second experiment, we added 8405 outliers (five times more than the inliers) in a $2 \times 2 \times 2$ cube containing the data.
- For the third experiment, we added Gaussian noise with variance 0.01 to the coordinates of all points.
- Finally, we embedded the perturbed data (and the outliers) in a 60-D space, before voting and nonlinear interpolation.

The noise-free and noisy input, as well as the generated points can be seen in Fig. 5.8. We computed the mean square error between the outputs generated by our method and Eq. 5.3 normalized by the variance of the noise-free data. The NMSE for all cases is reported in Table 5.10. Robustness against outliers is due to the fact that the inliers form a consistent surface

TABLE 5.10: Normalized MSE for the interpolated points of Eq. 5.3 under different noise conditions

EXPERIMENT	NMSE
Noise-free	0.0041
Outliers	0.0170
Outliers & N(0, 0.01)	0.0349
Outliers & N(0, 0.01) in 60-D	0.0241

and thus receive votes that support the correct local structure from other inliers. Outliers, on the other hand, are random and do not form any structure. They cast and receive inconsistent votes and therefore neither develop a preference for a certain manifold nor significantly disrupt the structure estimates at the inliers. They can be removed by simple thresholding since all their eigenvalues are small and almost equal. Note that performance in 60-D is actually better since the interference by outliers is reduced as the dimensionality of the space increases. Tensor voting is also robust against perturbation of the coordinates as long as its not biased to favor a certain direction. If the perturbation is zero-mean, its effects on individual votes are almost cancelled out, because they only contribute to the ball component of the accumulated tensor at each point, causing small errors in orientation estimation.

Results on Real Data. The final experiment is on real data taken from the University of California at Irvine Machine Learning Repository, which is available online at: http://www.ics.uci.edu/~mlearn/MLRepository.html. We used the "Auto-Mpg Database" that contains 392 samples of mileage per gallon (MPG) for automobiles as a function of seven discrete and continuous variables: number of cylinders, displacement, horsepower, weight, acceleration, model year and origin. Due to the large differences in the range of values for each variable, we re-scaled the data so that the ratio of maximum to minimum standard deviation of the variables was 10 : 1, instead of the original 1000 : 1. We randomly selected 314 samples, approximately 80% of the data, for training and 78 for testing. Since the variables are correlated, the samples do not form a manifold with seven degrees of freedom and do not cover the entire input domain. In fact, the estimated intrinsic dimensionality at each point by tensor voting ranges between one and three. After performing tensor voting on the 314 training samples, we estimate the MPG for the testing samples. We begin by finding the nearest neighbor in the 7-D input space for each testing sample and following a path on the estimated manifold in 8-D until the desired 7-D input coordinates are reached. The value of the output variable (MPG) when the input variables are equal to the query is the estimate returned by our method. The average error of our estimates with respect to the ground truth is 10.45%. Stable performance between 10.45% and 10.48% is achieved even as the average voting neighborhood ranges between 35 and 294 points. Considering that 314 points are hardly sufficient for inferring a description of a complex 8-D space, the performance of our algorithm is promising. In fact the average error we achieve is 2.67 miles per gallon, which we consider acceptable given the sparsity of the data.

5.6   DISCUSSION

We have presented an approach to manifold learning that offers certain advantages over the state of the art. In terms of dimensionality estimation, we are able to obtain accurate estimates at the point level. Moreover, since the dimensionality is found as the maximum gap in the eigenvalues of the tensor at each point, no thresholds are needed. In most other approaches, the dimensionality has to be provided, or, at best, an average intrinsic dimensionality is estimated for the entire dataset, as in [10, 13, 14, 41, 117].

Even though tensor voting on the surface looks similar to other local, instance-based learning algorithms that propagate information from point to point, the fact that the votes are tensors and not scalars allows them to convey considerably more information. The properties of the tensor representation, which can handle the simultaneous presence of multiple orientations, allow the reliable inference of the normal and tangent space at each point. In addition, tensor voting is very robust against outliers. This property holds in higher dimensions, where random noise is even more scattered.

It should also be noted that the votes attenuate with distance and curvature. This is a more intuitive formulation than using the K nearest neighbors with equal weights, since some of them may be too far, or belong to a different part of the structure. The only free parameter in our approach is σ, the scale of voting. Small values tend to preserve details better, while large values are more robust against noise. The scale can be selected automatically by randomly sampling a few points before voting and making sure that enough points are included in their voting neighborhoods. The number of points that can be considered sufficient is a function of the dimensionality of the space as well as the intrinsic dimensionality of the data. A full investigation of data sufficiency is among the objectives of our future research. Our results show that sensitivity with respect to scale is small, as shown in Tables 5.1, 5.3-5.5 and 5.9. The same can be observed in the results for the Auto-Mpg Dataset, where the error fluctuates by 0.03% as the average voting neighborhood ranges between 35 and 294 points.

Another important advantage of our approach is the absence of global computations, which enables us to process datasets with very large number of points. Memory requirements are in the order or $O(MN^2)$, where M is the number of points and N is the dimensionality of the input space. Time requirements are reasonably low, at $O(NM\log M)$. The basic operations are the sorting of the inputs which is $O(NM\log M)$, the retrieval of each point's neighbors from a k-d tree, which is $O(N \log M)$ for each of the M points and vote generation, which is linear in N for each voter and receiver pair. The eigen-decomposition of each tensor is $O(N^3 M)$, but this is not the bottleneck in practice since, for our method to be effective, the number of observations has to be considerably higher than the dimensionality of the space. Our algorithm fails when the available observations do not suffice to represent the manifold. This occurs, for instance, in the face with varying pose and illumination dataset of [109], where 698 instances represent a manifold in 4096-D. Graph-based methods are more successful in such situations. We do not view this fact as a serious limitation, since typical problems in machine learning are the over-abundance of data and the need for efficient processing of large datasets.

The novelty of our approach to manifold learning is that it is not based on dimensionality reduction, in the form of an embedding or mapping between a high and a low dimensional space. Instead, we perform tasks such as geodesic distance measurement and nonlinear interpolation in the input space. Experimental results show that we can perform these tasks in the presence of outlier noise at high accuracy, even without explicitly removing the outliers from the data. This is due to the fact that the accumulated tensors at the outliers do not develop any preference for a particular structure and do not outweigh the contributions of the inliers. In addition, outlier distribution remains random, since dimensionality reduction and an embedding to a lower-dimensional space are not attempted. This choice also broadens the range of datasets we can process. While isometric embeddings can be achieved for a certain class of manifolds, we are able to process non-flat manifolds and even non-manifolds. The last experiment of Section 5.4 demonstrates our ability to work with datasets of varying dimensionality or with intersecting manifolds. To the best of our knowledge, this is impossible with any other method. If dimensionality reduction is desired due to its considerable reduction in storage requirements, a dimensionality reduction method, such as [6, 10, 18, 82, 109, 117], can be used after tensor voting. The benefits of this process are in the form of noise robustness and smooth component identification, with respect to both dimensionality and orientation, via tensor voting followed by memory savings via dimensionality reduction.

We have also presented, in Section 5.5, a local nonparametric approach to function approximation that combines the advantages of local methods with the efficient representation and information propagation of tensor voting. Local function approximation methods are more flexible in the type of functions they can approximate, since the properties of the function are allowed to vary locally. Our approach, in particular, has no parameters, such as the number and type of local models to be used, that have to be selected, besides the scale of the voting field. Its drawback, in line with other local methods, is higher memory requirements, since the data have to be kept in memory. We have shown that we can process challenging examples from the literature under very adverse noise conditions. As shown in the example of Eq. 5.3, even when 80% of the samples are outliers and the inliers are corrupted by noise in the form of perturbation, we are still able to correctly predict unobserved outputs. Perturbation of the coordinates of the inliers by noise, especially when it is not zero-mean, can lead to errors in the estimates, especially at small scales. However, robustness against this type of noise is still rather high.

Our future research will focus on addressing the limitations of our current algorithm and extending its capabilities. In the area of function approximation, the issue of approximating functions with multiple branches for the same input value, which often appear in practical applications, has to be handled more rigorously. In addition, an interpolation mechanism that takes into account holes and boundaries should be implemented. We also intend to develop an online, incremental version of our approach, which will be able to process data as they are collected, instead of requiring the entire dataset to proceed. Potential applications of our work include challenging real problems, such as the study of direct and inverse kinematics. One can also view the proposed approach as learning data from a single class, which can serve as the groundwork for an approach for supervised and unsupervised classification.

CHAPTER 6

Boundary Inference

Many computer vision and machine learning applications require the reliable detection of boundaries, which is a particularly challenging problem in the presence of outliers and missing data. Here, we propose to address it by complementing the original tensor voting framework, which is limited to second-order properties, with first-order information. In this chapter, we define a first-order representation scheme that encodes the local distribution of neighbors around a token. We also define a mechanism to vote for the terminations of perceptual structures, such as the endpoints of curves. We take great care to ensure that the integration of first-order properties is in accordance with the philosophy of the framework. Namely, we maintain the capability to represent *all* structure types *simultaneously* and adhere to the principle of least commitment. What should be noted is that the first-order representation, even though it is complementary to the second-order one, can represent all boundary types. Some of the work presented in this chapter has been published in [112].

6.1 MOTIVATION

The second-order formulation of the original tensor voting framework, as shown in [60] and the previous chapters, is very effective for many perceptual organization problems. However, it is unable to detect terminations of open structures such as the endpoints of curves. It can be viewed as an excitatory process that facilitates grouping of the input data, and is able to extrapolate and infer dense salient structures. The integration of boundary inference, via first-order voting, provides a mechanism to inhibit the growth of the extracted structures. *Polarity vectors* are now associated with each token and encode the support the token receives for being a termination of a perceptual structure. The term *polarity* refers to the magnitude of the polarity vector. Polarity is large when the majority of the token's neighbors lie on the same side. The direction of the polarity vector indicates the direction of the inliers of the perceptual structure whose potential boundary is the token under consideration. The new representation exploits the essential property of boundaries to have all their neighbors, at least locally, on the same side of a half-space. As described in the remainder of the chapter, the voting scheme is identical to that of the second-order case.

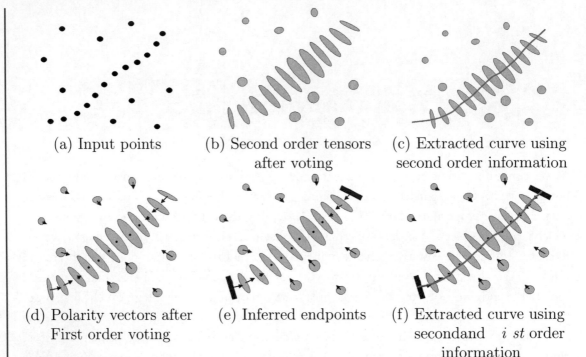

(a) Input points

(b) Second order tensors after voting

(c) Extracted curve using second order information

(d) Polarity vectors after First order voting

(e) Inferred endpoints

(f) Extracted curve using secondand $i st$ order information

FIGURE 6.1: Illustration of curve extraction with and without first-order voting. In the former case, even though the curve normals have been estimated correctly, there is no way other than heuristic thresholding to detect the endpoints and the curve extends beyond them, as seen in (c). On the other hand, when first-order information is available as in (d), the endpoints can be inferred as in (e) and the curve is terminated correctly as shown in (f).

A simple illustration of first-order voting can be seen in Fig. 6.1. The inputs are a few unoriented points that form a curve and some outliers (Fig. 6.1(a)), which are encoded as ball tensors. After second-order voting takes place, the inliers of the curves develop high stick saliency and the major axes of their tensors align with the normal orientation at each position (Fig. 6.1(b)). If we try to extract a continuous curve based on the strictly second-order formulation of [60], we begin marching from the most salient token and stop when stick saliency drops below a threshold. The resulting curve is shown in Fig. 6.1(c), where clearly it has grown beyond the endpoints, and is not consistent with the human interpretation of the input. If, instead, we perform first-order voting before curve extraction, we obtain a polarity vector at each token position (Fig. 6.1(d)). Nonmaximum suppression of polarity values can, then, be performed along the direction of the polarity vectors to detect the endpoints (Fig. 6.1(e)). Given the endpoints, the correct open curve can be extracted and is shown in Fig. 6.1(f).

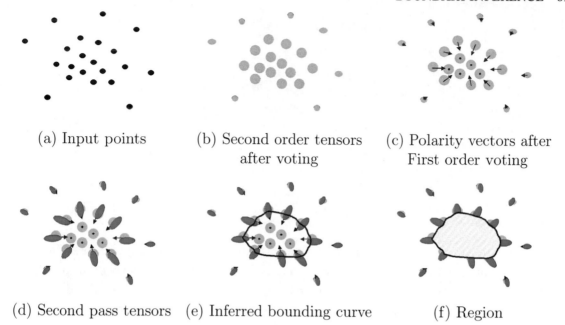

(a) Input points   (b) Second order tensors   (c) Polarity vectors after
                    after voting   First order voting

(d) Second pass tensors   (e) Inferred bounding curve   (f) Region

FIGURE 6.2: Illustration of region extraction. The inputs, in (a), are unoriented points encoded as ball tensors. The region inliers accumulate salient ball tensors after second-order voting, while the saliency of the outliers is smaller, as shown in (b). The boundaries accumulate consistent first-order votes and thus develop large polarity vectors shown in (c). These can be used as inputs for another pass of voting where points on salient boundary curves receive support from their neighbors, (d). Then a continuous curve can be extracted as in (e) and the desired region is the area enclosed by the bounding curve (f).

The same principle can be applied to region inference. We propose to infer regions via their boundaries, which in 2D are curves consisting of tokens with high polarity. Fig. 6.2 illustrates the process.

This chapter is organized as follows: Section 6.2 presents the first-order representation, voting, and voting fields, and shows how they are naturally derived from their second-order counterparts; Section 6.3 shows how the accumulated first- and second-order votes are analyzed to infer salient perceptual structures; Section 6.4 contains results on 2D and 3D synthetic, but challenging, datasets; and Section 6.5 concludes the chapter with a discussion on the contributions presented here and possible extensions.

6.2   FIRST-ORDER REPRESENTATION AND VOTING

The first-order information is conveyed by the polarity vector that encodes the likelihood of the token being on the boundary of a perceptual structure. Such boundaries in 2D are the endpoints of curves and the bounding curves of regions. In 3D, the possible types of boundaries are the

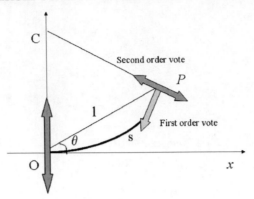

FIGURE 6.3: Second- and first-order votes cast by a stick tensor located at the origin FIX.

bounding surfaces of volumes, the bounding curves of surfaces, and the endpoints of curves. The direction of the polarity vector, in all cases, indicates the direction of the inliers of the perceptual structure whose potential boundary is the token under consideration. As before, the second-order tensor indicates the saliency of each type of perceptual structure the token belongs to and its preferred normal and tangent orientations. The two parts of the representation combined provide a much richer description of the tokens.

Now, we turn our attention to the generation of first-order votes. As shown in Fig. 6.3, the first-order vote cast by a unit stick tensor at the origin is *tangent* to the osculating circle, the smoothest path between the voter and receiver. Its magnitude, since nothing suggests otherwise, is equal to that of the second-order vote according to Eq. (2.2). The first-order voting field for a unit stick voter aligned with the z-axis is

$$\mathbf{S}_{\mathrm{FO}}(l, \theta, \sigma) = \mathrm{DF}(s, \kappa, \sigma) \begin{bmatrix} -\cos(2\theta) \\ -\sin(2\theta) \end{bmatrix}. \tag{6.1}$$

What should be noted is that tokens cast first- and second-order votes based on their second-order information only. This occurs because polarity vectors have to be initialized to zero since no assumption about structure terminations is available. Therefore, first-order votes are computed based on the second-order representation which can be initialized (in the form of ball tensors) even with no information other than the presence of a token. However, if first-order information was available, as a result of an endpoint detector for instance, it can be used for vote generation.

A simple illustration of how saliency and polarity values can be combined to infer curves and their endpoints in 2D appears in Fig. 6.4. The input consists of a set of collinear unoriented tokens which are encoded as ball tensors. The tokens cast votes to their neighbors and collect the votes cast to them. The accumulated curve saliency can be seen in Fig. 6.4(b), where the dashed

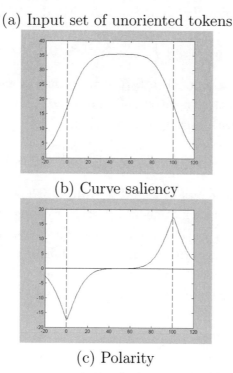

(a) Input set of unoriented tokens

(b) Curve saliency

(c) Polarity

FIGURE 6.4: Accumulated curve saliency and polarity for a simple input.

lines mark the limits of the input. The interior points of the curve receive more support and are more salient than those close to the endpoints. Since second-order voting is an excitatory process, locations beyond the endpoints are also compatible with the inferred line and receive consistent votes from the input tokens. Detection of the endpoints based on the second-order votes is virtually impossible since there is no systematic way of selecting a threshold that guarantees that the curve will not extend beyond the leftmost and rightmost input tokens. The accumulated polarity can be seen in Fig. 6.4(c). The endpoints appear clearly as maxima of polarity. The combination of saliency and polarity allows us to infer the curve and terminate it at the correct points.

Since in low-dimensional spaces it still makes sense to use precomputed voting fields, we briefly describe their computation here. The *2D first-order stick voting field* $\mathbf{S}_{FO}$ is a vector field, which at every position holds a vector that is equal in magnitude to the stick vote that exists at the same position in the fundamental second-order stick voting field, but is tangent to the smooth path between the voter and receiver instead of normal to it. The *first-order ball voting field* $\mathbf{B}_{FO}$ can be derived, as the second-order one, by integrating the contributions of a rotating stick tensor that casts first-order votes. The integration this time is approximated by *vector* instead of tensor addition. In 2D this is accomplished as follows,

$$\mathbf{B}_{FO}(P) = \int_0^{2\pi} R_\theta^{-1} \mathbf{S}_{FO}(R_\theta P) R_\theta^{-T} d\theta \tag{6.2}$$

where R_θ is the rotation matrix to align $\mathbf{S}_{FO}$ with $\hat{e}_1$, the eigenvector corresponding to the maximum eigenvalue (the stick component), of the rotating tensor at P. The 3D case is similar, but now the rotating stick tensor spans the unit sphere,

$$\mathbf{B}_{FO}(P) = \int_0^{\pi} \int_0^{\pi} R_{\theta\phi\psi}^{-1} \mathbf{S}_{FO}(R_{\theta\phi\psi} P) R_{\theta\phi\psi}^{-T} d\phi d\psi \Big|_{\theta=0} \tag{6.3}$$

where $R_{\theta\phi\psi}$ is the rotation matrix to align $\mathbf{S}_{FO}$ with $\hat{e}_1$, the eigenvector corresponding to the maximum eigenvalue (the stick component), of the rotating tensor at P, and θ, ϕ, ψ are rotation angles about the x-, y-, z-axis respectively. The *first-order plate voting field* $\mathbf{P}_{FO}$ is defined similarly. As in the second-order case, the voting fields are normalized so that the total energy is equal to that of the stick voting field. The norm of the vector votes is used as the measure of energy. Arbitrary tensors are decomposed into elementary tensors as in Eqs. (2.1) and (2.8), which cast votes using the appropriate fields. The vote of each component is weighted by the appropriate saliency value.

6.2.1   First-Order Voting in High Dimensions

In Chapter 4, we showed an approximation of vote generation that makes second-order vote generation from arbitrary tensors in arbitrary dimensions feasible. Here, we show how to derive the first-order votes in the same way as we derived the second-order ones. We consider three cases:

- *Stick voters* cast first-order votes that lie in the 2D plane defined by the position and orientation of the voter and the position of the receiver, are orthogonal to the second-order ones, and point toward the voter. The computation is identical to that of Eq. (6.1). Their magnitude is equal to that of the second-order vote.

- *Ball voters* cast first-order votes along $\vec{v}$, which is the vector parallel to the line connecting the voter and the receiver. The first-order votes point toward the voter and their magnitude is equal to the $N-1$ equal nonzero eigenvalues of the second-order vote.

- *Arbitrary voters with d equal eigenvalues*, unlike the second-order case, require merely one direct computation. Since the first-order vote has to be orthogonal to the normal space of the second-order vote, and also be tangent to the circular arc connecting the voter and receiver, it can be directly computed as the first-order vote cast by the stick voter $\vec{b}_1$ of Eq. (4.3.2). This vote satisfies all the requirements in terms of orientation and its magnitude decays with both the length and curvature of the arc, unless the receiver belongs to the tangent space of the voter and θ is zero.

6.3 VOTE ANALYSIS

Vote collection for the first-order case is performed by vector addition. The accumulated result is a vector whose direction points to a weighted center of mass from which votes are cast, and whose magnitude encodes polarity. Since the first-order votes are also weighted by the saliency of the voters and attenuate with distance and curvature, their vector sum points to the direction from which the most salient contributions were received. The accumulated first-order votes provide information that complements the accumulated second-order saliency information.

In 2D, a relatively low polarity value indicates that a token is in the interior of a curve or region; therefore surrounded by neighbors whose votes cancel each other out. On the other hand, high polarity indicates a token that is on or close to a boundary, thus receiving votes from only one side with respect to the boundary, at least locally. The correct boundaries can be extracted as local maxima of polarity along the direction of the polarity vector. Table 6.1 illustrates how tokens can be characterized using the collected first- and second-order information. In more detail, the cases that have to be considered are as follows:

- *Interior points of curves* can be found as local maxima of $\lambda_1 - \lambda_2$ along $\hat{e}_1$. In other words, they form the path of maximal curve saliency as one is marching along the curve's tangent starting from any token on the curve. Interior points receive first-order votes from both sides, which virtually cancel each other out. Their polarity is not a local maximum along the direction of the polarity vector.

- *Curve endpoints* have the same second-order properties, but they can be detected as local maxima of polarity along the direction of the polarity vector.

- *Interior points of regions* have high λ_2 values, as a result of the higher density of surrounding points.

- *Region boundaries* are region tokens that also have locally maximum polarity along the direction of the polarity vector. The latter is also orthogonal to the region boundaries.

- *Junctions* are isolated peaks of λ_2 as shown in Fig. 2.4. Their polarity depends on the type of junction. It is very small for X- and W-junctions, while it is high for L- and T-junctions.

- *Outliers* receive very little consistent support and have low saliency values compared to those of the inliers. Since they are in regions of low data density, their polarity values are also small due to the fact that votes attenuate with distance.

TABLE 6.1: Summary of First- and Second-Order Tensor Structure for Each Feature Type in 2D

2D FEATURE	SALIENCY	SECOND-ORDER TENSOR ORIENTATION	POLARITY	POLARITY VECTOR
Curve interior	Locally max $\lambda_1 - \lambda_2$ along $\hat{e}_1$	normal: $\hat{e}_1$	low	—
Curve endpoint	Locally max $\lambda_1 - \lambda_2$ along $\hat{e}_1$	normal: $\hat{e}_1$	Locally max along polarity vector	Parallel to $\hat{e}_2$
Region interior	High λ_2	—	Locally nonmax	—
Region boundary	High λ_2	—	Locally max along polarity vector	Normal to boundary
Junction	Local peak of λ_2	—	Depends on type	—
Outlier	Low	—	Low	—

In 3D, tokens can be classified according to the accumulated first- and second-order information according to Table 6.2. The accumulated second-order tensor is decomposed as in Section 2.3.3 and stick, plate, and ball saliencies are computed based on the eigenvalues. The tokens can be classified as follows:

- *Interior points of surfaces* have locally maximal $\lambda_1 - \lambda_2$ along the direction of the surface normal, $\hat{e}_1$. The surface in dense form can be extracted by the marching cubes algorithm [55] as the zero level of the first derivative of surface saliency. Unlike levels sets and marching cubes, in our case, the surface does not have to be closed. Interior points have nonmaximal polarity values.

- *Surface boundaries* have the same second-order properties as above, but also have locally maximal polarity along the direction of the polarity vector. The latter is orthogonal to the surface end-curve at the token's location.

- *Interior points of curves* have the same properties as in the 2D case, but now the appropriate saliency is given by $\lambda_2 - \lambda_3$. For a token to be on a curve its curve saliency has to be locally maximal on a plane normal to the curve spanned by $\hat{e}_1$ and $\hat{e}_2$, which are the two normals to the curve.

- *Curve endpoints* have the same properties as in the 2D case. The polarity vector in both cases is parallel to the tangent of the curve, the endpoints of which can be detected by their locally maximal polarity.

- *Interior points of regions* are characterized by the same properties as in the 2D case. The differences are that, in 3D, ball saliency is given by λ_3 and that regions are volumes.

- *Region boundaries* are volume boundaries in 3D and the analysis is the same as in the 2D case. The polarity vector is orthogonal to the bounding surface of the volume.

- *Junctions* are local peaks of λ_3. Their polarity depends on the type of junction as in the 2D case. Convex vertices of polyhedra have polarity vectors that point to the interior of the polyhedron, while the polarity vector points to the outside at concave vertices.

- *Outliers*, as before, receive very little consistent support and have low saliency and polarity values.

In ND, vote analysis is a direct generalization of the low-dimensional cases, with the only difference being that $N + 1$ structure types are possible in an ND space. First-order information is utilized to infer the boundaries of structures, the type of which is given after analysis of the second-order tensors.

TABLE 6.2: Summary of the First- and Second-Order Tensor Structure for Each Feature Type in 3D

3D FEATURE	SALIENCY	SECOND-ORDER TENSOR ORIENTATION	POLARITY	POLARITY VECTOR
Surface interior	Locally max $\lambda_1 - \lambda_2$ along $\hat{e}_1$	Normal: $\hat{e}_1$	Low	—
Surface end-curve	Locally max $\lambda_1 - \lambda_2$ along $\hat{e}_1$	Normal: $\hat{e}_1$	Locally max $\lambda_1 - \lambda_2$ along polarity Vector	Orthogonal to $\hat{e}_1$ and end-curve
Curve interior	Locally max $\lambda_2 - \lambda_3$ on $\hat{e}_1, \hat{e}_2$ plane	Tangent: $\hat{e}_3$	Low	—
Curve endpoint	Locally max $\lambda_2 - \lambda_3$ on $\hat{e}_1, \hat{e}_2$ plane	Tangent: $\hat{e}_3$	Locally max along polarity vector	Parallel to $\hat{e}_3$
Region interior	High λ_3	—	Low	—
Region boundary	High λ_3	—	Locally max along polarity vector	Normal to bounding surface
Junction	Local peak of λ_3	—	Depends on type	—
Outlier	Low	—	Low	—

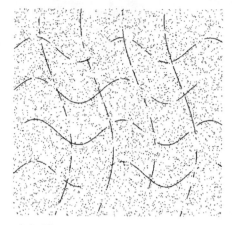
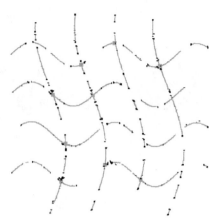

(a) Noisy unoriented data set   (b) Extracted curves, endpoints and junctions

FIGURE 6.5: Curve, endpoint, and junction extraction on a noisy dataset with sinusoidal curves.

6.4   RESULTS USING FIRST-ORDER INFORMATION

In this section, we present results on synthetic datasets corrupted by noise. In all cases, processing begins with unoriented tokens, which are encoded as ball tensors. The capability to proceed with oriented or unoriented tokens, or a mixture of both, is a feature of tensor voting not shared by many other approaches.

Results in 2D.   Fig. 6.5(a) shows a dataset that contains a number of fragmented sinusoidal curves represented by unoriented points contaminated by a large number of outliers. All points are encoded as ball tensors. Fig. 6.5(b) shows the output after tensor voting. Curve inliers are colored gray, endpoints black, while junctions appear as gray squares. The noise has been removed, and the curve segments have been correctly detected and their endpoints and junctions labeled.

More results are illustrated in Fig. 6.6, which demonstrates the simultaneous detection of a region and a curve, as well as their terminations. The curve is inferred even as it goes through the region, since curve saliency is still locally maximal due to the higher density of curve inliers compared to the region inliers. The curve by itself is shown in Fig. 6.6(c) for clarity.

Results in 3D.   Given a noisy set of points that belong to a 3D region, as in Fig. 6.7(a), we infer volume boundaries as local maxima of polarity along the direction of the polarity vector. In terms of second-order tensors, volume inliers are characterized by a dominant ball component, since they collect second-order votes from all directions in 3D. The same holds for tokens close to the boundaries, since second-order votes are a function of orientation but not direction. The

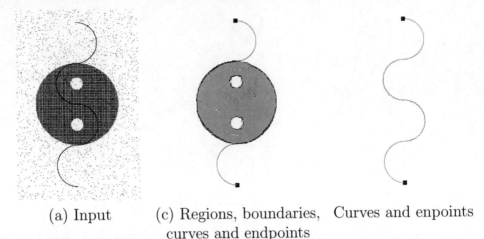

(a) Input   (c) Regions, boundaries,   Curves and enpoints
                       curves and endpoints

FIGURE 6.6: Regions and curves from a noisy point set. Region boundaries and curve endpoints are marked in black. The curve is shown by itself on the right to aid visualization.

bounding surface of a 3D region can be extracted by the modified surface marching algorithm [103, 104] as the maximal isosurface of polarity along the normal direction, indicated by the polarity vectors. Fig. 6.7(a) depicts two solid generalized cylinders with different parabolic sweep functions. The cylinders are generated by a uniform random distribution of unoriented points in their interior, while the noise is also uniformly distributed but with a lower density. After tensor voting, volume inliers are detected due to their high ball saliency and low polarity, while volume boundaries are detected due to their high ball saliency and polarity. The polarity vectors are normal to the bounding surface of the cylinders. Using the detected boundaries as

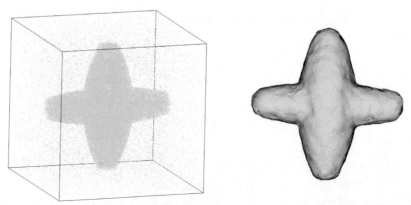

(a) Noisy unoriented data set   (b) Dense bounding surface

FIGURE 6.7: Region inlier and boundary detection in 3D.

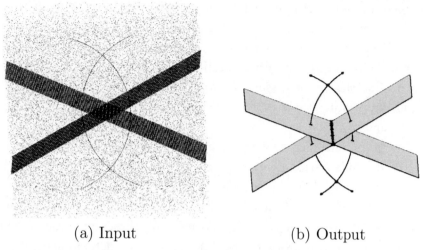

(a) Input   (b) Output

FIGURE 6.8: Results on simultaneous inference of multiple types of structures. (a) Unoriented dataset that consists of two intersecting planes, two intersecting curves, and random outliers. (b) Output after voting. Outliers have been rejected due to very low saliency. Surface inliers are marked in gray, curves and boundaries in black. Curve endpoints and junctions have been enlarged.

input we perform dense voting and extract the bounding surface in continuous form in Fig. 6.7(b).

We close this section with an example of simultaneous inference of surfaces, curves, surface intersections, junctions, surface boundaries, and curve endpoints, which is presented in Fig. 6.8. The simultaneous inference of *all* types of structures and the interaction among them is a feature that can only be found in the tensor voting framework. Methods based on optimization would have to be run once for each structure type while their output at the intersection of different types, such as a curve–surface intersection, would most likely be unclear.

6.5   DISCUSSION

In this chapter, we have presented a critical augmentation to the tensor voting framework. It deals with the fundamental smoothness versus discontinuities dilemma that occurs in most nontrivial perceptual organization scenarios. Many perceptual organization approaches operate either as grouping or as segmentation processes. We believe that both grouping *and* segmentation must be performed in order to tackle challenging problems. In both cases, boundaries play a critical part.

It is worth noting that the integration of first-order information does not violate the principles on which the original framework is founded. The approach is model free and makes no assumptions about the input other than that salient perceptual structures are generated by the good alignment of tokens. All processing is local within the neighborhood of each token, and local changes of the input result in local changes of the output. Grouping and segmentation are still performed in a soft way, in the sense that the first- and second-order votes from a token to another express the degree of affinity between the tokens, but there is no hard labeling. Hard labeling can be the final stage of processing, if it is required by a specific application. In addition, first- and second-order voting can be performed simultaneously, resulting in a small increase in computational complexity.

In addition to the capability to detect terminations in itself, the work presented in this chapter serves as the foundation for more complex perceptual organization problems. In Chapter 7 , we propose a novel approach for figure completion in which endpoints and labeled junctions play a critical role. The inference of these keypoints and the classification of junctions as T, L, X, etc. is the first step in that direction. This is possible only through the use of the first-order augmentation presented in this chapter. Then, the possible modal and amodal completions supported by each type of keypoint can be examined to produce hypotheses for completion.

Finally, the inferred terminations can serve as interfaces between structures in a multiscale processing scheme. This is especially useful for datasets where data density varies considerably. Even though it can be viewed as a case of amodal figure completion in 2D, multiscale processing, using the inferred terminations as inputs to the next scale, can be easily extended to higher dimensions. Encouraging results on medical imaging using this approach can be found in [112].

CHAPTER 7

Figure Completion

In this chapter, we address the issues associated with figure completion, a fundamental perceptual grouping task. Endpoints and junctions, which can be detected based on their first-order properties, play a critical role in contour completion by the human visual system and should be an integral part of a computational process that attempts to emulate human perception. A significant body of evidence in the psychology literature points to two types of completion: modal (or orthogonal) and amodal (or parallel). We present a methodology within the tensor voting framework which implements both types of completion and integrates a fully automatic decision-making mechanism for selecting between them. Our approach was initially published in [67] and proceeds directly from tokens or binary image input, infers descriptions in terms of overlapping layers, and labels junctions as T, L, and endpoints. It is based on first- and second-order tensor voting, which facilitates the propagation of local support among tokens. The addition of first-order information to the original framework is crucial, since it makes the inference of endpoints and the labeling of junctions possible. We illustrate the approach on several classical inputs, producing interpretations consistent with those of the human visual system.

7.1 INTRODUCTION

Figure completion is an important component of image understanding that has received a lot of attention from both the computational vision and the neuroscience community over the past few decades. While we do not claim that our approach is biologically plausible, the human visual system serves as the paradigm, since even a small fraction of its performance still evades all attempts to emulate it. In this chapter, we show that the interpretations we infer are consistent with human perception even in difficult cases. Consider the fragmented contour of Fig. 7.1(a) which supports amodal completion of the half circle, as in Fig. 7.1(b). Note that completion stops at the two endpoints, marked *A* and *B*. Now consider Fig. 7.1(c), which is known as the Ehrenstein stimulus. This supports modal completion and produces a strong and unique perception of an illusory circle (Fig. 7.1(c)). Note that the outer circle is not perceived, most probably because it would require a concave occluder, which is unlikely and thus not preferred by the human visual system. What makes this input interesting to us is that both types of completion

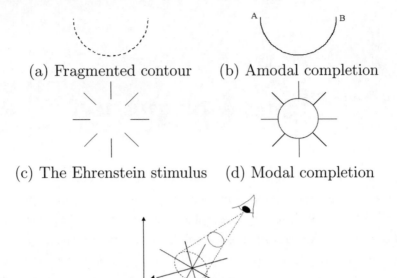

(a) Fragmented contour   (b) Amodal completion

(c) The Ehrenstein stimulus   (d) Modal completion

(e) Layered interpretation of the Ehrenstein stimulus

FIGURE 7.1: Amodal and modal completions.

are supported by the data: modal completion of the occluding disk and amodal completion of the occluded lines. We aim at inferring a description which is consistent with human perception that produces a layered interpretation of the image, with a white disk occluding a set of black lines on a white background (Fig. 7.1(e)).

An aspect of computer vision that has not been solved is the selection between modal and amodal completion. Most researchers assume that the type of completion to be performed is known in advance. For instance, modal completion in most cases starts only when the endpoints and the orientation of completion are provided as inputs. We aim at inferring the endpoints and junctions from unoriented data, and then, automatically making decisions whether further completion is supported by the data, and which type of completion should occur. The tensor voting framework is well suited for this task since it allows the integrated inference of curves, junctions, regions, and terminations. The latter is possible after the incorporation of first-order information into the framework.

Classical approaches to the inference of salient contours include the work of Grossberg, Mingolla, and Todorovic [22, 23]. They developed the *boundary contour system* (BCS) and the *feature contour system* (FCS) that can group fragmented and even illusory edges to form closed boundaries and regions by feature cooperation and competition in a neural network. Gove et al. [21] present a paper within the BCS/FCS framework that specifically addresses the perception of illusory contours. In the work of Heitger and von der Heydt [29] elementary curves are grouped into contours via convolution with a set of orientation-selective kernels, whose responses decay with distance and difference in orientation. Mechanisms for both parallel and orthogonal grouping based on keypoints such as junctions and line ends are also proposed. Williams and Thornber [121] extend the *stochastic completion fields* framework of Williams and Jacobs [119] to address modal and amodal completion. They describe the inference of closed illusory contours from position and orientation constraints that can be derived from line terminations.

Our work differs from other approaches in that we can proceed from unoriented, unlabeled data and simultaneously infer curves, junctions, regions, and boundaries. Moreover, we propose an *automatic* mechanism for making decisions between modal and amodal completion without having to know the type of completion in advance.

7.2   OVERVIEW OF THE APPROACH

The input to our algorithm is a set of tokens in a two-dimensional space. The tokens indicate the presence of a primitive that potentially belongs to a larger configuration and can be generated by a process such as an edge detector in the form of a bank of filters. The output is a layered description of the image in terms of salient curves, junctions, and regions. Our first goal is to detect salient groupings based on the support tokens receive from their neighbors. The amount of support from one token to another is in the form of a first- and a second-order vote whose properties depend on proximity, collinearity, and cocurvilinearity. Since the representation of each token can simultaneously encode its behavior as a curvel, a junction, or a region inlier, tokens do not have to be classified prematurely, and decisions are made when enough information is available. Here, we take advantage of first-order properties to infer endpoints and region boundaries, as well as label junctions.

The novelty of our work is the mechanism for modal and amodal completion using the endpoints and junctions inferred at the previous stage. Endpoints, T-junctions, and L-junctions offer two possibilities for completion. Either completion along the endpoint's tangent or the T-junction's stem if a corresponding keypoint with compatible orientation exists within a certain vicinity (amodal completion), or completion along the orthogonal orientation of an endpoint, the bar of a T-junction or an edge of an L-junction (modal completion). To facilitate modal and amodal completion, two new vote generation mechanisms are defined based on standard vote generation. The decision whether completion is supported by the data is made by examining the support for both options at every keypoint. If at least one more keypoint supports modal or amodal completion, the current keypoint is labeled as one that supports completion, and the appropriate voting field is used. If both modal and amodal completion are supported, both are further pursued and a layered description is inferred. The process and the fields are described in more detail in Section 7.4.

7.3   TENSOR VOTING ON LOW LEVEL INPUTS

Since natural images, besides presenting insurmountable difficulties in the extraction of meaningful descriptions in terms of object outlines, do not typically exhibit modal completion, we use synthetic data to demonstrate our approach. Edge detection and junction detection are considerably easier for these synthetic images, but the focus here is on the higher level processing stage, where completion occurs. In the examples presented here, black pixels are encoded as unit ball tensors. Since we want to collect saliency information everywhere, the remaining grid positions are initialized with null tensors. Dense saliency maps allow us to perform nonmaximal suppression for the inference of salient structures.

In addition to curves and keypoints, regions are also inferred based on their high λ_2 and enclosure by region boundaries. The latter can be inferred after nonmaximum suppression with respect to polarity along the direction of the polarity vector. If regions are inferred in the data, an additional step is required. The boundaries of the inferred regions, which are detected based on their high polarity and ball saliency, participate in a round of tensor voting along with the curvels and junctions to infer region–curve intersections. The set of endpoints and junctions that is passed on to the next module is determined after analyzing the saliency maps after this additional stage.

7.4   COMPLETION

Now that endpoints and junctions have been inferred and labeled, we need to consider how they interact to produce figure completion. There are two types of completion: modal and amodal. In *amodal* completion, endpoints extend along their tangent and T-junctions along their stem. This case corresponds to the completion of an *occluded* surface behind the occluder. In *modal* completion connections occur along the bar of T-junctions; orthogonally to the tangent of endpoints, which are interpreted as low-contrast T-junctions; or along one of the edges of L-junctions, which are also interpreted as low-contrast T-junctions. Modal completion is the completion of the *occluding* surface on top of the occluded ones. Amodal completion and modal completion are termed, respectively, parallel and orthogonal grouping by Heitger and von der Heydt [29].

Our framework makes automatic decisions on which type of completion occurs. Either type has to be supported by at least one other keypoint within a certain range, which is larger than that of the first stage. For instance, amodal completion from an endpoint has to be supported by another endpoint with similar orientation and opposite polarity. Then, tensor voting is performed to infer a contour connecting the two endpoints, as with the endpoints of the curve segments of Fig. 7.1(a). Amodal completion between pairs of endpoints produces the contour of Fig. 7.1(b).

New voting fields, based on the original second-order stick voting field, have to be defined to facilitate the propagation of information from the voting endpoints and junctions. The

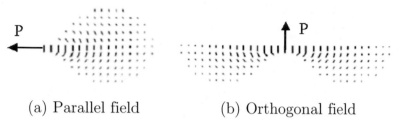

(a) Parallel field   (b) Orthogonal field

FIGURE 7.2: Voting fields used for amodal and modal completion. P is the polarity vector at the voting endpoint or junction. Voting is allowed only toward the opposite half-plane.

orientation and saliency of the votes are the same as in Eqs. (2.2) and (2.3), but the fields are one sided. This is because amodal completion can only occur along one direction: away from the curve that was inferred at the first stage or in the direction that would make the T-junction an X-junction. Modal completion is possible only toward the direction that results in a convex illusory contour. This is due to a strong preference in the human visual system for convex modal completions [25]. Even though the two fields are orthogonal, the polarity vector, in both cases, indicates the direction opposite to completion. Fig. 7.2 shows the fields used for these cases. Before voting, the following cases have to be considered for each keypoint:

- There is no possible modal or amodal continuation due to the absence of other keypoints that support either option. In this case endpoints are just terminations, like A and B in Fig. 7.1(b).

- Amodal completion is supported by another keypoint of the same type with similar orientation and opposite polarity, while modal is not (Fig. 7.1(a)). Endpoints and stems of T-junctions cast votes with the parallel field along their tangent.

- Modal completion is possible, but amodal is not (Fig. 7.3). Support in this case has to come from keypoints with similar curve orientation and polarity. Completion occurs orthogonally to the tangent of endpoints or along the bars of T-junctions and edges of L-junctions, using the orthogonal field.

- Both types of completion are possible (Fig. 7.1(c)). In this case, the modal completion is perceived as occluding the amodal completion. In the case of the Ehrenstein stimulus, a disk is perceived to occlude the crossing lines (Fig. 7.1(e)).

Once the above cases have been examined, the keypoints can be labeled with respect to whether they support completion or not. The appropriate field is used according to the type of completion. Analysis of the votes is performed and curves, junctions, and endpoints are inferred as in the previous section. Now, junctions can be fully labeled according to their cardinality and

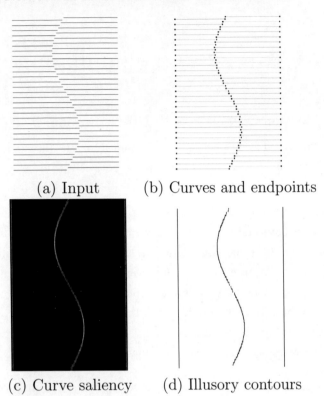

FIGURE 7.3: Modal completion. Starting from unoriented inputs, endpoints are detected and used as inputs for modal completion.

polarity. Polarity helps to discriminate between X- and W-junctions for instance, since the former have very low polarity while the latter have high polarity.

7.5 EXPERIMENTAL RESULTS

We now present experimental results in a variety of examples. A small scale is used for the original data and a large scale (typically 20 to 30 times larger) is used to infer completions at the second stage, where the distance between tokens is considerably larger.

Illusory Contour Without Depth Ordering. This is an example of the formation of an open illusory contour that does not induce the perception of a depth ordering. In that sense it is similar to Fig. 8 of [93]. The input consists of a set of unoriented tokens that form line segments and can be seen in Fig. 7.3(a). The curves and endpoints detected after a first pass of tensor voting can be seen in Fig. 7.3(b). The endpoints are marked in black and their inferred normals are orthogonal to the segments. Three illusory contours can be inferred after voting

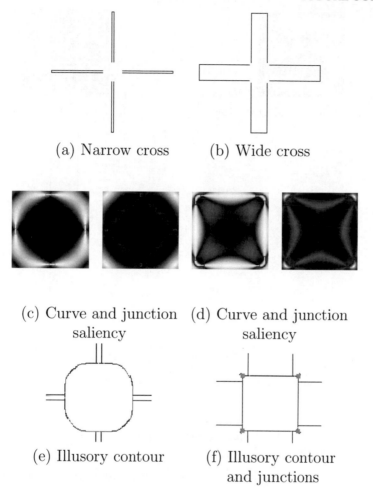

(a) Narrow cross   (b) Wide cross

(c) Curve and junction   (d) Curve and junction
saliency   saliency

(e) Illusory contour   (f) Illusory contour
                          and junctions

FIGURE 7.4: Koffka crosses. Inputs, saliency maps, and inferred contours and junctions (marked as squares).

using the orthogonal field of Fig. 7.2(b). Curve saliency and the illusory contours can be seen in Figs. 7.3(c) and (d). Higher intensity corresponds to higher saliency, as in all saliency maps shown in this document. The contour is still inferred, even though its convexity changes, since locally endpoints from either the left or the right side form a convex contour and propagate votes that support the entire sinusoidal contour.

Koffka Crosses. An interesting perceptual phenomenon is the Koffka crosses [121]. The perceived illusory contour changes from a circle to a square as the arms of the cross become wider. Two examples of Koffka crosses can be seen in Figs. 7.4(a) and (b). The black pixels of these images are encoded as unoriented ball tensors and the endpoints of the segments are extracted

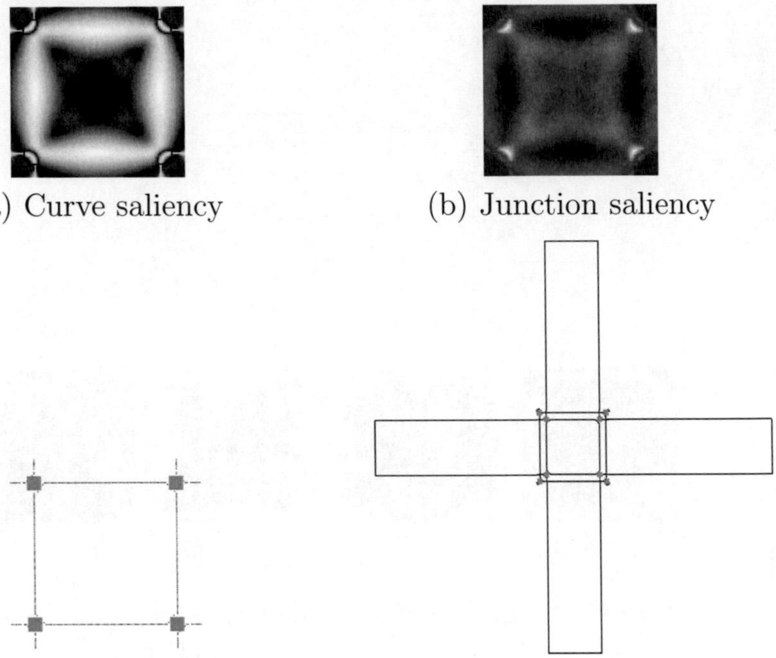

(a) Curve saliency

(b) Junction saliency

(c) Illusory contours and junctions

(d) All completions

FIGURE 7.5: Amodal contour completion and illustration of all possible completions, including the occluded ones.

as before. Modal completion is possible, since the endpoints can be hypothesized as T-junctions with zero contrast bars, and voting is performed using the orthogonal field. Curve and junction saliencies are shown in Figs. 7.4(c) and (d). Note that the saliencies in each map are normalized independently so that white corresponds to the maximum and black to the minimum. The maximum junction saliency is 90.4% of the maximum curve saliency for the wide cross and only 9.8% of the maximum curve saliency for the narrow cross, where no junctions are inferred. Figs. 7.4(e) and (f) show the inferred modal completion. Intermediate widths of the arms produce intermediate shapes of the illusory contour, such as rounded squares, which are consistent with human perception.

The case of amodal completion from the detected endpoints to compatible endpoints must also be considered. The parallel voting field of Fig. 7.2(a) should be used in this case. Figs. 7.5(a) and (b) show the curve and junction saliencies for the wide Koffka cross of Fig. 7.4(b). Four contours that connect corresponding endpoints and four X-junctions are inferred (Fig. 7.5(c)). This interpretation is also consistent with human perception, which is a layered interpretation of the scene that consists of a white square, occluding a cross, on top of a white background.

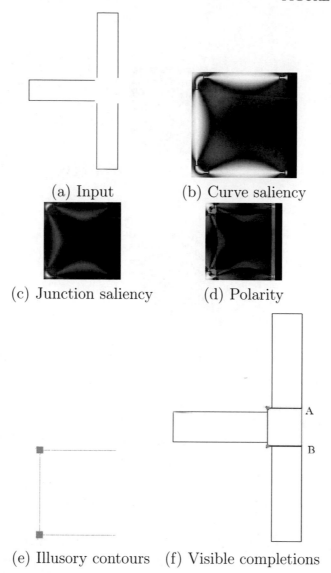

(a) Input   (b) Curve saliency

(c) Junction saliency   (d) Polarity

(e) Illusory contours   (f) Visible completions

FIGURE 7.6: Modal completion for the three-armed cross.

We also performed an experiment with one arm of the wide cross missing. The input can be seen in Fig. 7.6(a). Voting with the orthogonal field produces the saliency and polarity maps seen in Figs. 7.6(b) and (d). The polarity map has four local maxima: at the two L-junctions and at two of the six voting endpoints. The inferred description consists of three straight contours, two L-junctions and two endpoints at the wrongly hypothesized T-junctions, A and B, which based on polarity can now be correctly labeled as L-junctions. Additional modal completion is now possible starting from the L-junctions that results in contour AB in Fig. 7.6(f).

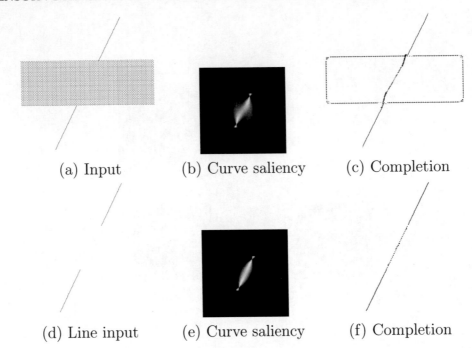

(a) Input   (b) Curve saliency   (c) Completion

(d) Line input   (e) Curve saliency   (f) Completion

FIGURE 7.7: Possible explanation of the illusion that straight lines appear bent when passing through a region.

Poggendorff Illusion.   In the final example of this chapter, we attempt to explain an illusion that occurs when a straight line does not appear perfectly straight when it passes through a region. The input can be seen in Fig. 7.7(a) and consists of unoriented points that form a line and a region. The region boundaries are inferred based on their high λ_2 and high polarity and are used as inputs to a second pass of tensor voting along with the curvels. After the second pass, four L-junctions and two T-junctions are inferred. There is no support for completion based on the L-junctions so they do not contribute any further. The T-junctions should be treated carefully in this case. The orientation of the stems is locally unreliable, since as one approaches the junction, curve saliency decreases almost to zero while junction saliency increases. Even if orientation θ of the line farther away from the stem is used, some uncertainty remains. On the other hand, the bar and the stem of a T-junction are expected to be orthogonal [25]. Combining these two sources of evidence, we can set the orientation of the stem equal to $\alpha 90° + (1 - \alpha)\theta$. Voting from the two T-junctions using the parallel field, for a range of values of α, produces a curve saliency map like that of Fig. 7.7(b) where completion is not straight and an inflection point always exists in the contour (Fig. 7.7(c)). This is one possible explanation for the illusion. Also, note that the illusion does not occur if the line is orthogonal to the region boundaries. The

(a) The Kanizsa triangle   (b) Completion of the triangle   (c) Completion of the disks

FIGURE 7.8: The Kanizsa triangle and two possible interpretations.

saliency map in the absence of the region is that shown in Fig. 7.7(e) where straight continuation occurs, as expected.

7.6   DISCUSSION

In this chapter, we have presented an approach within the tensor voting framework that allows us to infer richer descriptions from single images. We begin by inferring the endpoints of curves and the boundaries of regions, and assigning preliminary labels to junctions. Moreover, we have addressed figure completion of both modal and amodal type in an automatic and integrated way. Our framework does not require a priori knowledge of the type of completion that has to be performed, but can infer it from the data. We demonstrated how it can be applied to the interpretation of line drawings that exhibit complex perceptual phenomena.

Consistently with the body of work on tensor voting, we have avoided premature decisions based on local operators. For instance, we do not classify a location as an L-junction just because a corner detector produces a strong response. Instead, we only assign a preliminary label based on the results of first- and second-order voting, which for an L-junction have to produce high ball saliency and high polarity. The final labeling occurs only after the completion possibilities have been examined.

The results are encouraging. However, there are still numerous issues that need to be addressed, even in simple line drawings. Consider, for instance, Fig. 7.8(a) that depicts the Kanizsa triangle [40]. It contains six L-junctions and each junction supports two possibilities for completion. Either completion along the straight edge, which produces the triangle of Fig. 7.8(b), or completion along the circular edge, which produces the three disks seen in 7.8(c). This example is an excellent demonstration of a scenario that cannot be handled by the current state of our research. Moreover, note that making the decision on one vertex of the triangle affects the other two vertices. As demonstrated by numerous visual illusions, drawings of impossible objects and in [93], for instance, locally consistent perceptions that are globally impossible are accepted by the human visual system. Therefore, the decision on one vertex does not automatically resolve the other two. What is clear, however, is that the computer vision, psychology, and neuroscience literature provide abundant examples for which a more sophisticated decision-making mechanism than the one presented here is needed.

In this chapter, we have also scratched the surface of inferring hierarchical descriptions. Typically, processing occurs in two stages: in the first stage, tensor voting is performed on the original low level tokens, while in the second stage, completion based on the previously inferred structures is performed. There are three processing stages in case regions are present in the dataset. Then, region boundaries are inferred in the first stage and interact with other tokens at the second stage. Completion now occurs at the third stage. More complicated scenarios may include more stages. It is reasonable to assume that scale increases from stage to stage, as the distances between the "active" points increase. A more systematic investigation of the role of scale in this context is also required. It is possible that the interpretation of certain inputs changes as the scale of voting varies.

CHAPTER 8

Conclusions

In the previous chapters, we described both a general perceptual organization approach as well as its application to a number of computer vision and machine learning problems. The cornerstone of our work is the tensor voting framework, which provides a powerful and flexible way to infer the saliency of structures formed by elementary primitives. The primitives may differ from problem to problem, but the philosophy behind the manner in which we address them is the same. In all cases, we arrive at solutions which receive maximal support from the primitives as the most coherent and smooth structures. Throughout this work, we strove to maintain the desired properties that we described in the introduction. The approach should be local, data driven, unsupervised, robust to noise, and able to represent all structure types simultaneously. These principles make our approach general and flexible, while allowing us to incorporate problem-specific constraints as, for instance, uniqueness for stereo.

While we have shown promising results, which in many cases compare favorably to the state of the art in a variety of fields, we feel that there is still a lot of work to be done within the framework. This work ranges from the 2D case, where the inference of integrated description in terms of edges, junctions, and regions has received a lot of attention from the research community, but is far from being considered solved, to the ND machine learning case. The research presented in Chapters 4 and 5 has only scratched the surface of the capabilities of our approach and will serve as the groundwork for research in domains that include pattern recognition, classification, data mining, and kinematics. Unlike competing approaches, tensor voting scales well as the number of samples increases since it involves only local computations. This property is crucial in a world where information is generated and transmitted a lot faster than it can be processed. Our experiments have demonstrated that we can attain excellent performance levels given sufficient samples, and the latter abound in many cases.

References

[1] S. Arya, D. M. Mount, N. S. Netanyahu, R. Silverman and A. Y. Wu, "An optimal algorithm for approximate nearest neighbor searching," *J. ACM*, Vol. 45, pp. 891–923, 1998. doi:10.1145/293347.293348

[2] C. G. Atkeson, A. W. Moore and S. Schaal, "Locally weighted learning," *Artif. Intell. Rev.*, Vol. 11(1–5), pp. 11–73, 1997. doi:10.1023/A:1006559212014

[3] A. R. Barron, "Universal approximation bounds for superpositions of a sigmoidal function," *IEEE Trans. Inf. Theory*, Vol. 39(3), pp. 930–945, 1993. doi:10.1109/18.256500

[4] P. N. Belhumeur, "A Bayesian-approach to binocular stereopsis," *Int. J. Comput. Vis.*, Vol. 19(3), pp. 237–260, August 1996. doi:10.1007/BF00055146

[5] P. N. Belhumeur and D. Mumford, "A Bayesian treatment of the stereo correspondence problem using half-occluded regions," in *Int. Conf. on Computer Vision and Pattern Recognition*, 1992, pp. 506–512.

[6] M. Belkin and P. Niyogi, "Laplacian eigenmaps for dimensionality reduction and data representation," *Neural Comput.*, Vol. 15(6), pp. 1373–1396, 2003. doi:10.1162/089976603321780317

[7] S. Birchfield and C. Tomasi, "A pixel dissimilarity measure that is insensitive to image sampling," *IEEE Trans. Pattern Anal. Mach. Intell.*, Vol. 20(4), pp. 401–406, April 1998.

[8] S. Birchfield and C. Tomasi, "Multiway cut for stereo and motion with slanted surfaces," in *Int. Conf. on Computer Vision*, 1999, pp. 489–495.

[9] A. F. Bobick and S. S. Intille, "Large occlusion stereo," *Int. J Comput. Vis.*, Vol. 33(3), pp. 1–20, Sept. 1999.

[10] M. Brand, "Charting a manifold," in *Advances in Neural Information Processing Systems*, Vol. 15. Cambridge, MA: MIT Press, 2003, pp. 961–968.

[11] L. Breiman, "Hinging hyperplanes for regression, classification, and function approximation," *IEEE Trans. Inf. Theory*, Vol. 39(3), pp. 999–1013, 1993. doi:10.1109/18.256506

[12] M. Z. Brown, D. Burschka and G. D. Hager, "Advances in computational stereo," *IEEE Trans. Pattern Anal. Mach. Intell.*, Vol. 25(8), pp. 993–1008, Aug. 2003. doi:10.1109/TPAMI.2003.1217603

[13] J. Bruske and G. Sommer, "Intrinsic dimensionality estimation with optimally topology preserving maps," *IEEE Trans. Pattern Anal. Mach. Intell.*, Vol. 20(5), pp. 572–575, May 1998. doi:10.1109/34.682189

[14] J. Costa and A. O. Hero, "Geodesic entropic graphs for dimension and entropy estimation in manifold learning," *IEEE Trans. Signal Process.*, Vol. 52(8), pp. 2210–2221, Aug. 2004. doi:10.1109/TSP.2004.831130

[15] T. Cox and M. Cox, *Multidimensional Scaling.* London: Chapman and Hall, 1994.

[16] V. de Silva and J. B. Tenenbaum, "Global versus local methods in nonlinear dimensionality reduction," in *Advances in Neural Information Processing Systems*, Vol. 15. Cambridge, MA: MIT Press, 2003, pp. 705–712.

[17] J. Dolan and E. M. Riseman, "Computing curvilinear structure by token-based grouping," in *Int. Conf. on Computer Vision and Pattern Recognition*, 1992, pp. 264–270.

[18] D. Donoho and C. Grimes, "Hessian eigenmaps: new tools for nonlinear dimensionality reduction," in *Proceedings of National Academy of Science*, 2003, pp. 5591–5596.

[19] O. D. Faugeras and R. Keriven, "Variational principles, surface evolution, PDEs, level set methods, and the stereo problem," *IEEE Trans. Image Process.*, Vol. 7(3), pp. 336–344, March 1998. doi:10.1109/83.661183

[20] D. Geiger, B. Ladendorf, and A. Yuille, "Occlusions and binocular stereo," *Int. J. Comput. Vis.*, Vol. 14(3), pp. 211–226, April 1995.

[21] A. Gove, S. Grossberg and E. Mingolla, "Brightness perception, illusory contours, and corticogeniculate feedback," *Vis. Neurosci.*, Vol. 12, pp. 1027–1052, 1995.

[22] S. Grossberg and E. Mingolla, "Neural dynamics of form perception: Boundary completion," *Psychol. Rev.*, Vol. 92(2), pp. 173–211, 1985.

[23] S. Grossberg and D. Todorovic, "Neural dynamics of 1-d and 2-d brightness perception: A unified model of classical and recent phenomena," *Percept. Psychophys.*, Vol. 43, pp. 723–742, 1988.

[24] G. Guy, "Inference of multiple curves and surfaces from sparse data," Ph.D. Thesis, University of Southern California, 1995.

[25] G. Guy and G. Medioni, "Inferring global perceptual contours from local features," *Int. J. Comput. Vis.*, Vol. 20(1/2), pp. 113–133, 1996.

[26] G. Guy and G. Medioni, "Inference of surfaces, 3d curves, and junctions from sparse, noisy, 3d data," *IEEE Trans. Pattern Anal. Mach. Intell.*, Vol. 19(11), pp. 1265–1277, Nov. 1997.

[27] R. I. Hartley and A. Zisserman, *Multiple View Geometry in Computer Vision.* Cambridge: Cambridge University Press, 2000.

[28] X. He and P. Niyogi, "Locality preserving projections," in *Advances in Neural Information Processing Systems*, Vol. 16, S. Thrun, L. Saul and B. Schölkopf, Eds. Cambridge, MA: MIT Press, 2004.

[29]   F. Heitger and R. von der Heydt, "A computational model of neural contour processing: Figure-ground segregation and illusory contours," in *Int. Conf. on Computer Vision*, 1993, pp. 32–40.

[30]   W. Hoff and N. Ahuja, "Surfaces from stereo: Integrating feature matching, disparity estimation, and contour detection," *IEEE Trans. Pattern Anal. Mach. Intell.*, Vol. 11(2), pp. 121–136, Feb. 1989. doi:10.1109/34.16709

[31]   S. S. Intille and A. F. Bobick, "Disparity-space images and large occlusion stereo," in *European Conf. on Computer Vision*, 1994, pp. B: 179–186.

[32]   H. Ishikawa and D. Geiger, "Occlusions, discontinuities, and epipolar lines in stereo," in *European Conf. on Computer Vision*, 1998, pp. I: 232–248.

[33]   L. Itti and P. Baldi, "A principled approach to detecting surprising events in video," in *Int. Conf. on Computer Vision and Pattern Recognition*, Jun 2005.

[34]   D. W. Jacobs, "Robust and efficient detection of salient convex groups," *IEEE Trans. Pattern Anal. Mach. Intell.*, Vol. 18(1), pp. 23–37, Jan. 1996. doi:10.1109/34.476008

[35]   J. Jia and C. K. Tang, "Image repairing: Robust image synthesis by adaptive nd tensor voting," in *Int. Conf. on Computer Vision and Pattern Recognition*, 2003, pp. I: 643–650.

[36]   J. Jia and C. K. Tang, "Inference of segmented color and texture description by tensor voting," *IEEE Trans. Pattern Anal. Mach. Intell.*, Vol. 26(6), pp. 771–786, 2004. doi:10.1109/TPAMI.2004.10

[37]   I. T. Jolliffe, *Principal Component Analysis*. New York: Springer, 1986.

[38]   T. Kanade and M. Okutomi, "A stereo matching algorithm with an adaptive window: Theory and experiment," *IEEE Trans. Pattern Anal. Mach. Intell.*, Vol. 16(9), pp. 920–932, Sept. 1994. doi:10.1109/34.310690

[39]   J. Kang, I. Cohen and G. Medioni, "Continuous multi-views tracking using tensor voting," in *IEEE Workshop on Motion and Video Computing*, 2002, pp. 181–186.

[40]   G. Kanizsa, *Organization in Vision*. New York: Praeger, 1979.

[41]   B. Kégl, "Intrinsic dimension estimation using packing numbers," in *Advances in Neural Information Processing Systems*, Vol. 15. Cambridge, MA: MIT Press, 2005, pp. 681–688.

[42]   K. Koffka, *Principles of Gestalt Psychology*. New York: Harcourt/Brace, 1935.

[43]   W. Köhler, "Physical gestalten," in *A Source Book of Gestalt Psychology (1950)*, W. D. Ellis, Ed. New York: Harcourt/Brace, 1920, pp. 17–54.

[44]   V. Kolmogorov and R. Zabih, "Computing visual correspondence with occlusions via graph cuts," in *Int. Conf. on Computer Vision*, 2001, pp. II: 508–515.

[45]   V. Kolmogorov and R. Zabih, "Multi-camera scene reconstruction via graph cuts," in *European Conf. on Computer Vision*, pp. III: 82–96, 2002.

[46]   P. Kornprobst and G. Medioni, "Tracking segmented objects using tensor voting," in *Int. Conf. on Computer Vision and Pattern Recognition*, 2000, pp. II: 118–125.

[47]   K. N. Kutulakos and S. M. Seitz, "A theory of shape by space carving," *Int. J. Comput. Vis.*, Vol. 38(3), pp. 199–218, July 2000.

[48]   S. Lawrence, A. C. Tsoi and A. D. Back, "Function approximation with neural networks and local methods: Bias, variance and smoothness," in *Australian Conference on Neural Networks*, 1996, pp. 16–21.

[49]   M. S. Lee, "Tensor voting for salient feature inference in computer vision," Ph.D. Thesis, University of Southern California, 1998.

[50]   M. S. Lee and G. Medioni, "Inferring segmented surface description from stereo data," in *Int. Conf. on Computer Vision and Pattern Recognition*, 1998, pp. 346–352.

[51]   M. S. Lee, G. Medioni, and P. Mordohai, "Inference of segmented overlapping surfaces from binocular stereo," *IEEE Trans. Pattern Anal. Mach. Intell.*, Vol. 24(6), pp. 824–837, June 2002. doi:10.1109/TPAMI.2002.1008388

[52]   E. Levina and P. Bickel, "Maximum likelihood estimation of intrinsic dimension," in *Advances in Neural Information Processing Systems*, Vol. 17. Cambridge, MA: MIT Press, 2005, pp. 777–784.

[53]   Z. Li, "A neural model of contour integration in the primary visual cortex," *Neural Comput.*, Vol. 10, pp. 903–940, 1998. doi:10.1162/089976698300017557

[54]   M. H. Lin and C. Tomasi, "Surfaces with occlusions from layered stereo," in *Int. Conf. on Computer Vision and Pattern Recognition*, 2003, pp. I: 710–717.

[55]   W. E. Lorensen and H. E. Cline, "Marching cubes: A high resolution 3d surface reconstruction algorithm," *Comput. Graph.*, Vol. 21(4), pp. 163–169, 1987. doi:10.1016/0097-8493(87)90030-6

[56]   D. G. Lowe, *Perceptual Organization and Visual Recognition*. Dordrecht: Kluwer, June 1985.

[57]   A. Luo and H. Burkhardt, "An intensity-based cooperative bidirectional stereo matching with simultaneous detection of discontinuities and occlusions," *Int. J. Comput. Vis.*, Vol. 15(3), pp. 171–188, July 1995.

[58]   D. Marr, *Vision*. San Francisco: Freeman, 1982.

[59]   D. Marr and T. A. Poggio, "Cooperative computation of stereo disparity," *Science*, Vol. 194(4262), pp. 283–287, Oct. 1976.

[60]   G. Medioni, M. S. Lee, and C. K. Tang, *A Computational Framework for Segmentation and Grouping*. New York: Elsevier, 2000.

[61]   S. Mitaim and B. Kosko, "The shape of fuzzy sets in adaptive function approximation," *IEEE Trans. Fuzzy Syst.*, Vol. 9(4), pp. 637–656, 2001. doi:10.1109/91.940974

[62]   T. M. Mitchell, *Machine Learning*. New York: McGraw-Hill, 1997.

[63] R. Mohan and R. Nevatia, "Perceptual organization for scene segmentation and description," *IEEE Trans. Pattern Anal. Mach. Intell.*, Vol. 14(6), pp. 616–635, June 1992. doi:10.1109/34.141553

[64] P. Mordohai, "A perceptual organization approach for figure completion, binocular and multiple-view stereo and machine learning using tensor voting," Ph.D. Thesis, University of Southern California, 2005.

[65] P. Mordohai and G. Medioni, "Perceptual grouping for multiple view stereo using tensor voting," in *Int. Conf. on Pattern Recognition*, 2002, pp. III: 639–644.

[66] P. Mordohai and G. Medioni, "Dense multiple view stereo with general camera placement using tensor voting," in *2nd Int. Symp. on 3-D Data Processing, Visualization and Transmission*, 2004, pp. 725–732.

[67] P. Mordohai and G. Medioni, "Junction inference and classification for figure completion using tensor voting," in *4th Workshop on Perceptual Organization in Computer Vision*, 2004, p. 56.

[68] P. Mordohai and G. Medioni, "Stereo using monocular cues within the tensor voting framework," in *European Conf. on Computer Vision*, 2004, pp. 588–601.

[69] P. Mordohai and G. Medioni, "Unsupervised dimensionality estimation and manifold learning in high-dimensional spaces by tensor voting," in *Int. Joint Conf. on Artificial Intelligence*, 2005, to be published.

[70] P. Mordohai and G. Medioni, "Stereo using monocular cues within the tensor voting framework," *IEEE Trans. Pattern Anal. Mach. Intell.*, 2006, to be published.

[71] H. Neummann and E. Mingolla, "Computational neural models of spatial integration in perceptual grouping," in *From Fragments to Objects: Grouping and Segmentation in Vision*, T. F. Shipley and P. J. Kellman, Eds. Berkeley, CA: Peachpit, 2001, pp. 353–400.

[72] M. Nicolescu and G. Medioni, "4-d voting for matching, densification and segmentation into motion layers," in *Int. Conf. on Pattern Recognition*, 2002, pp. III: 303–308.

[73] M. Nicolescu and G. Medioni, "Perceptual grouping from motion cues using tensor voting in 4-d," in *European Conf. on Computer Vision*, 2002, pp. III: 423–428.

[74] M. Nicolescu and G. Medioni, "Layered 4d representation and voting for grouping from motion," *IEEE Trans. Pattern Anal. Mach. Intell.*, Vol. 25(4), pp. 492–501, April 2003.

[75] M. Nicolescu and G. Medioni. "Motion segmentation with accurate boundaries—a tensor voting approach," in *Int. Conf. on Computer Vision and Pattern Recognition*, 2003, pp. I: 382–389.

[76] A. S. Ogale and Y. Aloimonos, "Stereo correspondence with slanted surfaces: Critical implications of horizontal slant," in *Int. Conf. on Computer Vision and Pattern Recognition*, 2004, pp. I: 568–573.

[77] Y. Ohta and T. Kanade, "Stereo by intra- and inter-scanline search using dynamic programming," *IEEE Trans. Pattern Anal. Mach. Intell.*, Vol. 7(2), pp. 139–154, March 1985.

[78] S. Osher and R. P. Fedkiw, *The Level Set Method and Dynamic Implicit Surfaces*. Berlin: Springer, 2002.

[79] P. Parent and S. W. Zucker, "Trace inference, curvature consistency, and curve detection," *IEEE Trans. Pattern Anal. Mach. Intell.*, Vol. 11(8), pp. 823–839, Aug. 1989. doi:10.1109/34.31445

[80] T. Poggio and F. Girosi, "Networks for approximation and learning," *Proc. IEEE*, Vol. 78(9), pp. 1481–1497, 1990. doi:10.1109/5.58326

[81] S. B. Pollard, J. E. W. Mayhew and J. P. Frisby, "Pmf: A stereo correspondence algorithm using a disparity gradient limit," *Perception*, Vol. 14, pp. 449–470, 1985.

[82] S. T. Roweis and L. K. Saul, "Nonlinear dimensionality reduction by locally linear embedding," *Science*, Vol. 290, pp. 2323–2326, 2000. doi:10.1126/science.290.5500.2323

[83] S. Roy and I. J. Cox, "A maximum-flow formulation of the n-camera stereo correspondence problem," in *Int. Conf. on Computer Vision*, 1998, pp. 492–499.

[84] S. Russell and P. Norvig, *Artificial Intelligence: A Modern Approach*. Englewood Cliffs, NJ: Prentice-Hall, 2003.

[85] A. Saha, C. L. Wu and D. S. Tang, "Approximation, dimension reduction, and non-convex optimization using linear superpositions of Gaussians," *IEEE Trans. Comput.*, Vol. 42(10), pp. 1222–1233, 1993. doi:10.1109/12.257708

[86] P. T. Sander and S. W. Zucker, "Inferring surface trace and differential structure from 3-d images," *IEEE Trans. Pattern Anal. Mach. Intell.*, Vol. 12(9), pp. 833–854, Sept. 1990. doi:10.1109/34.57680

[87] T. D. Sanger, "A tree-structured algorithm for reducing computation in networks with separable basis functions," *Neural Comput.*, Vol. 3(1), pp. 67–78, 1991.

[88] R. Sara, "Finding the largest unambiguous component of stereo matching," in *European Conf. on Computer Vision*, 2002, pp. III: 900–914.

[89] S. Sarkar and K. L. Boyer, "A computational structure for preattentive perceptual organization: Graphical enumeration and voting methods," *IEEE Trans. Syst. Man Cybern.*, Vol. 24, pp. 246–267, 1994. doi:10.1109/21.281424

[90] L. K. Saul and S. T. Roweis, "Think globally, fit locally: unsupervised learning of low dimensional manifolds," *J. Mach. Learn. Res.*, Vol. 4, pp. 119–155, 2003. doi:10.1162/153244304322972667

[91] E. Saund, "Symbolic construction of a 2-d scale-space image," *IEEE Trans. Pattern Anal. Mach. Intell.*, Vol. 12(8), pp. 817–830, Aug. 1990. doi:10.1109/34.57672

[92] E. Saund, "Labeling of curvilinear structure across scales by token grouping," in *Int. Conf. on Computer Vision and Pattern Recognition*, 1992, pp. 257–263.

[93] E. Saund, "Perceptual organization of occluding contours of opaque surfaces," *Comput. Vis. Image Underst.*, Vol. 76(1), pp. 70–82, Oct. 1999.

[94] S. Schaal and C. G. Atkeson, "Constructive incremental learning from only local information," *Neural Comput.*, Vol. 10(8), pp. 2047–2084, 1998. doi:10.1162/089976698300016963

[95] D. Scharstein and R. Szeliski, "Stereo matching with nonlinear diffusion," *Int. J. Comput. Vis.*, Vol. 28(2), pp. 155–174, 1998. doi:10.1023/A:1008015117424

[96] D. Scharstein and R. Szeliski, "A taxonomy and evaluation of dense two-frame stereo correspondence algorithms," *Int. J. Comput. Vis.*, Vol. 47(1–3), pp. 7–42, April 2002.

[97] D. Scharstein and R. Szeliski, "High-accuracy stereo depth maps using structured light," in *Int. Conf. on Computer Vision and Pattern Recognition*, 2003, pp. I: 195–202.

[98] B. Sch'olkopf, A. J. Smola and K.-R. Müller, "Nonlinear component analysis as a kernel eigenvalue problem," *Neural Comput.*, Vol. 10(5), pp. 1299–1319, 1998. doi:10.1162/089976698300017467

[99] A. Shashua and S. Ullman, "Structural saliency: The detection of globally salient structures using a locally connected network," in *Int. Conf. on Computer Vision*, 1988, pp. 321–327.

[100] J. Sun, Y. Li, S. B. Kang and H. Y. Shum, "Symmetric stereo matching for occlusion handling," in *Int. Conf. on Computer Vision and Pattern Recognition*, 2005, pp. II: 399–406.

[101] J. Sun, N. N. Zheng and H. Y. Shum, "Stereo matching using belief propagation," *IEEE Trans. Pattern Anal. Mach. Intell.*, Vol. 25(7), pp. 787–800, July 2003.

[102] R. Szeliski and D. Scharstein, "Symmetric sub-pixel stereo matching," in *European Conf. on Computer Vision*, 2002, pp. II: 525–540.

[103] C. K. Tang, "Tensor voting in computer vision, visualization, and higher dimensional inferences," Ph.D. Thesis, University of Southern California, 2000.

[104] C. K. Tang and G. Medioni, "Inference of integrated surface, curve, and junction descriptions from sparse 3d data," *IEEE Trans. Pattern Anal. Mach. Intell.*, Vol. 20(11), pp. 1206–1223, Nov. 1998.

[105] C. K. Tang, G. Medioni and M. S. Lee, "Epipolar geometry estimation by tensor voting in 8d," in *Int. Conf. on Computer Vision*, 1999, pp. 502–509.

[106] C. K. Tang, G. Medioni and M. S. Lee, "N-dimensional tensor voting and application to epipolar geometry estimation," *IEEE Trans. Pattern Anal. Mach. Intell.*, Vol. 23(8), pp. 829–844, Aug. 2001. doi:10.1109/34.946987

[107] M. F. Tappen and W. T. Freeman, "Comparison of graph cuts with belief propagation for stereo, using identical MRF parameters," in *Int. Conf. on Computer Vision*, 2003, pp. 900–907.

[108] Y. W. Teh and S. Roweis, "Automatic alignment of local representations," in *Advances in Neural Information Processing Systems*, Vol. 15. Cambridge, MA: MIT Press, 2003, pp. 841–848.

[109] J. B. Tenenbaum, V. de Silva and J. C. Langford, "A global geometric framework for nonlinear dimensionality reduction," *Science*, Vol. 290, pp. 2319–2323, 2000. doi:10.1126/science.290.5500.2319

[110] W. S. Tong, C. K. Tang and G. Medioni, "Epipolar geometry estimation for non-static scenes by 4d tensor voting," in *Int. Conf. on Computer Vision and Pattern Recognition*, 2001, pp. I: 926–933.

[111] W. S. Tong, C. K. Tang and G. Medioni, "Simultaneous two-view epipolar geometry estimation and motion segmentation by 4d tensor voting," *IEEE Trans. Pattern Anal. Mach. Intell.*, Vol. 26(9), pp. 1167–1184, Sept. 2004. doi:10.1109/TPAMI.2004.72

[112] W. S. Tong, C. K. Tang, P. Mordohai and G. Medioni, "First order augmentation to tensor voting for boundary inference and multiscale analysis in 3d," *IEEE Trans. Pattern Anal. Mach. Intell.*, Vol. 26(5), pp. 594–611, May 2004. doi:10.1109/TPAMI.2004.1273934

[113] O. Veksler, "Fast variable window for stereo correspondence using integral images," in *Int. Conf. on Computer Vision and Pattern Recognition*, 2003, pp. I: 556–561.

[114] S. Vijayakumar, A. D'Souza, T. Shibata, J. Conradt and S. Schaal, "Statistical learning for humanoid robots," *Auton. Robots*, Vol. 12(1), pp. 59–72, 2002.

[115] S. Vijayakumar and S. Schaal, "Locally weighted projection regression: An o(n) algorithm for incremental real time learning in high dimensional space," in *Int. Conf. on Machine Learning*, 2000, pp. I: 288–293.

[116] J. Wang, Z. Zhang and H. Zha, "Adaptive manifold learning," in *Advances in Neural Information Processing Systems*, Vol. 17, L. K. Saul, Y. Weiss and L. Bottou, Eds. Cambridge, MA: MIT Press, 2005.

[117] K. Q. Weinberger and L. K. Saul, "Unsupervised learning of image manifolds by semidefinite programming," in *Proc. Int. Conf. on Computer Vision and Pattern Recognition*, 2004, pp. II: 988–995.

[118] M. Wertheimer, "Laws of organization in perceptual forms," in *Psycologische Forschung, Translation by W. Ellis, A Source Book of Gestalt Psychology (1938)* Vol. 4. Cambridge, MA: MIT Press 1923, pp. 301–350.

[119] L. R. Williams and D. W. Jacobs, "Stochastic completion fields: A neural model of illusory contour shape and salience," *Neural Comput.*, Vol. 9(4), pp. 837–858, 1997. doi:10.1162/neco.1997.9.4.837

[120] L. R. Williams and K. K. Thornber, "A comparison of measures for detecting natural shapes in cluttered backgrounds," *Int. J. Comput. Vis.*, Vol. 34(2–3), pp. 81–96, Aug. 1999. doi:10.1023/A:1008187804026

[121] L. R. Williams and K. K. Thornber, "Orientation, scale, and discontinuity as emergent properties of illusory contour shape," *Neural Comput.*, Vol. 13(8), pp. 1683–1711, 2001. doi:10.1162/08997660152469305

[122] L. Xu, M. I. Jordan and G. E. Hinton, "An alternative model for mixtures of experts," in *Advances in Neural Information Processing Systems*, Vol. 7, G. Tesauro, D. S. Touretzky and T. K. Leen, Eds. Cambridge, MA: MIT Press, 1995, pp. 633–640.

[123] S. C. Yen and L. H. Finkel, "Extraction of perceptually salient contours by striate cortical networks," *Vis. Res.*, Vol. 38(5), pp. 719–741, 1998.

[124] A. J. Yezzi and S. Soatto, "Stereoscopic segmentation," in *Int. Conf. on Computer Vision*, 2001, pp. I: 59–66.

[125] Y. Zhang and C. Kambhamettu, "Stereo matching with segmentation-based coopera-tion," in *European Conf. on Computer Vision*, 2002, pp. II: 556–571.

[126] Z. Zhang and Y. Shan, "A progressive scheme for stereo matching," in *Lecture Notes on Computer Science*, Vol. 2018. Berlin: Springer, 2001, pp. 68–85.

[127] Z. Zhang and H. Zha, "Principal manifolds and nonlinear dimension reduction via local tangent space alignment," *SIAM J. Sci. Comput.*, Vol. 26(1), pp. 313–338, 2004.

[128] Z. Y. Zhang, "Determining the epipolar geometry and its uncertainty: A review," *Int. J. Comput. Vis.*, Vol. 27(2), pp. 161–195, March 1998. doi:10.1023/A:1007941100561

[129] C. L. Zitnick and T. Kanade, "A cooperative algorithm for stereo matching and occlusion detection," *IEEE Trans. Pattern Anal. Mach. Intell.*, Vol. 22(7), pp. 675–684, July 2000.

Author Biographies

Philippos Mordohai   received his Diploma in Electrical and Computer Engineering from the Aristotle University of Thessaloniki, Greece, in 1998. He also received the MS and PhD degrees both in Electrical Engineering from the University of Southern California, Los Angeles, in 2000 and 2005, respectively. He is currently a postdoctoral research associate at the Department of Computer Science of the University of North Carolina in Chapel Hill. His doctoral dissertation work focused on the development of perceptual organization approaches for computer vision and machine learning problems. The topics he has worked on include feature inference in images, figure completion, binocular and multiple-view stereo, instance-based learning, dimensionality estimation, and function approximation. His current research is on the 3D reconstruction of urban environments from multiple video cameras mounted on a moving vehicle. Dr Mordohai is a member of the IEEE and the IEEE Computer Society, reviewer for the Transactions on Pattern Analysis and Machine Intelligence and the Transactions on Neural Networks. He served as chair of local organization for the Third International Symposium on 3D Data Processing, Visualization and Transmission that was held in Chapel Hill in 2006.

Gérard Medioni   received the Diplôme d' Ingénieur Civil from the Ecole Nationale Supérieure des Télécommunications, Paris, France, in 1977, and the MS and PhD degrees in Computer Science from the University of Southern California, Los Angeles, in 1980 and 1983, respectively. He has been with the University of Southern California (USC) in Los Angeles, since 1983, where he is currently a professor of Computer Science and Electrical Engineering, codirector of the Computer Vision Laboratory, and chairman of the Computer Science Department. He was a visiting scientist at INRIA Sophia Antipolis in 1993 and Chief Technical Officer of Geometrix, Inc. during his sabbatical leave in 2000. His research interests cover a broad spectrum of the computer vision field and he has studied techniques for edge detection, perceptual grouping, shape description, stereo analysis, range image understanding, image to map correspondence, object recognition, and image sequence analysis. He has published more than 100 papers in conference proceedings and journals. Dr Medioni is a Fellow of the IEEE and a Fellow of the IAPR. He has served on the program committees of many major vision conferences and was program chairman of the 1991 IEEE Computer Vision and Pattern Recognition Conference in Maui, program cochairman of the 1995 IEEE Symposium on Computer Vision held in Coral Gables, Florida, general cochair of the 1997

IEEE Computer Vision and Pattern Recognition Conference in Puerto Rico, program cochair of the 1998 International Conference on Pattern Recognition held in Brisbane, Australia, and general cochairman of the 2001 IEEE Computer Vision and Pattern Recognition Conference in Kauai. Professor Medioni is on the editorial board of the Pattern Recognition and Image Analysis journal and the International Journal of Computer Vision and one of the North American editors for the Image and Vision Computing journal.

ted in the United States
ker & Taylor Publisher Services